Future R&D Environments
A Report for the
National Institute of Standards and Technology

Committee on Future Environments for the
National Institute of Standards and Technology

Division on Engineering and Physical Sciences

National Research Council

NATIONAL ACADEMY PRESS
Washington, D.C.

NATIONAL ACADEMY PRESS • 2101 Constitution Avenue, N.W. • Washington, DC 20418

NOTICE: The project that is the subject of this report was approved by the Governing Board of the National Research Council, whose members are drawn from the councils of the National Academy of Sciences, the National Academy of Engineering, and the Institute of Medicine. The members of the committee responsible for the report were chosen for their special competences and with regard for appropriate balance.

This study was supported by Contract No. 50SBNBOC1003 between the National Academy of Sciences and the National Institute of Standards and Technology. Any opinions, findings, conclusions, or recommendations expressed in this publication are those of the author(s) and do not necessarily reflect the views of the organizations or agencies that provided support for the project.

International Standard Book Number 0-309-08336-2

Additional copies of this report are available from National Academy Press, 2101 Constitution Avenue, N.W., Lockbox 285, Washington, DC 20055; (800) 624-6242 or (202) 334-3313 (in the Washington metropolitan area); Internet, http://www.nap.edu

Copyright 2002 by the National Academy of Sciences. All rights reserved.

Printed in the United States of America

THE NATIONAL ACADEMIES

National Academy of Sciences
National Academy of Engineering
Institute of Medicine
National Research Council

The **National Academy of Sciences** is a private, nonprofit, self-perpetuating society of distinguished scholars engaged in scientific and engineering research, dedicated to the furtherance of science and technology and to their use for the general welfare. Upon the authority of the charter granted to it by the Congress in 1863, the Academy has a mandate that requires it to advise the federal government on scientific and technical matters. Dr. Bruce M. Alberts is president of the National Academy of Sciences.

The **National Academy of Engineering** was established in 1964, under the charter of the National Academy of Sciences, as a parallel organization of outstanding engineers. It is autonomous in its administration and in the selection of its members, sharing with the National Academy of Sciences the responsibility for advising the federal government. The National Academy of Engineering also sponsors engineering programs aimed at meeting national needs, encourages education and research, and recognizes the superior achievements of engineers. Dr. Wm. A. Wulf is president of the National Academy of Engineering.

The **Institute of Medicine** was established in 1970 by the National Academy of Sciences to secure the services of eminent members of appropriate professions in the examination of policy matters pertaining to the health of the public. The Institute acts under the responsibility given to the National Academy of Sciences by its congressional charter to be an adviser to the federal government and, upon its own initiative, to identify issues of medical care, research, and education. Dr. Kenneth I. Shine is president of the Institute of Medicine.

The **National Research Council** was organized by the National Academy of Sciences in 1916 to associate the broad community of science and technology with the Academy's purposes of furthering knowledge and advising the federal government. Functioning in accordance with general policies determined by the Academy, the Council has become the principal operating agency of both the National Academy of Sciences and the National Academy of Engineering in providing services to the government, the public, and the scientific and engineering communities. The Council is administered jointly by both Academies and the Institute of Medicine. Dr. Bruce M. Alberts and Dr. Wm. A. Wulf are chairman and vice chairman, respectively, of the National Research Council.

COMMITTEE ON FUTURE ENVIRONMENTS FOR THE NATIONAL INSTITUTE OF STANDARDS AND TECHNOLOGY

KENNETH H. KELLER, University of Minnesota, *Chair*
MILTON CHANG, iNCUBiC, LLC
WILLIAM E. COYNE, 3M Corporation
JAMES W. DALLY, University of Maryland, College Park
CHARLES P. DeLISI, Boston University
C. WILLIAM GEAR, NEC Research Institute, Inc.
ROY LEVIN, Microsoft Corporation
RICHARD L. POPP, Stanford University School of Medicine
NATHAN ROSENBERG, Stanford University
THOMAS A. SAPONAS, Agilent Technologies

Staff

NORMAN METZGER, Study Director
MICHAEL McGEARY, Consultant (Acting Study Director, November 20, 2001–March 5, 2002)
STEPHEN A. MERRILL, Executive Director, Board on Science, Technology, and Economic Policy
MARIA P. JONES, Senior Project Assistant

DIVISION ON ENGINEERING AND PHYSICAL SCIENCES

WILLIAM A. WULF, National Academy of Engineering, *Chair*
WILLIAM F. BALLHAUS, JR., The Aerospace Corporation
PETER M. BANKS, XR Ventures, LLC
SHIRLEY CHIANG, University of California at Davis
MARSHALL H. COHEN, California Institute of Technology
INGRID DAUBECHIES, Princeton University
SAMUEL H. FULLER, Analog Devices, Inc.
PAUL H. GILBERT, Parsons Brinckerhoff International, Inc.
WESLEY T. HUNTRESS, JR., Carnegie Institution
TREVOR O. JONES, BIOMEC, Inc.
NANCY G. LEVESON, Massachusetts Institute of Technology
CORA B. MARRETT, University of Massachusetts at Amherst
ROBERT M. NEREM, Georgia Institute of Technology
JANET L. NORWOOD, Former Commissioner, U.S. Bureau of Labor Statistics
LAWRENCE T. PAPAY, Science Applications International Corporation
WILLIAM H. PRESS, Los Alamos National Laboratory
ROBERT J. SPINRAD, Xerox PARC (retired)
BARRY M. TROST, Stanford University
JAMES C. WILLIAMS, Ohio State University

PETER D. BLAIR, Executive Director

Preface

In September 2000, the deputy director of the National Institute of Standards and Technology (NIST) asked the National Research Council to perform the following task:

> The Commission on Physical Sciences, Mathematics, and Applications [which as of January 1, 2001, became part of the Division on Engineering and Physical Sciences] will examine forces and trends over the next 5 to 10 years pertinent to NIST's mission. The basis will be the judgments of a well-rounded committee, supported by a facilitated workshop probing a range of possible trends and forces in science and technology, the economy, industry, and other areas that NIST should consider in its future planning. The examination will be complemented by a review of recent presentations at the Academies' symposia on frontiers in science and engineering. Neither a "roadmap" nor projections of specific future outcomes will be provided.

The aim was to assist NIST in planning future programs in fulfillment of its stated role of "strengthening the U.S. economy and improving the quality of life by working with industry to develop and apply technology, measurements, and standards." Against this, the National Research Council was asked to set out a range of possible directions that science and its technological applications may take, influenced by forces and trends in the economy and in industrial management and strength, and, of course, not least by current frontiers in science and technology. NIST did not ask the National Research Council to provide specific predictions or projections. Nor did it request guidance on how NIST management might translate possible future directions identified by the committee into specific programs and organization.

Accordingly, the Committee on Future Environments for the National Institute of Standards and Technology sought neither to predict nor to project, but rather to set out a range of possible futures for the direction of science and technology. It approached the task in several complementary ways. First, it broke the task into examining "push," "pull," and "contextual" factors. "Push" gathered together the committee's judgments on possible "futures" for a set of scientific and technical fields, focusing on biology and medicine, materials, and information technology. "Pull" focused on societal demand factors—the economic, social, environmental, and political needs and sensitivities that would promote or inhibit research and development in certain areas of science and technology, as well as innovations based on that R&D. Under contextual factors, the committee considered a set of issues such as changes in the organization and support of R&D in both the public and private sectors, educational goals of students and methods of delivering education, and patterns of investment by the private sector, all of which might be expected to change the process by which ideas move from research to product. While obviously this classification of factors is somewhat arbitrary, the committee nevertheless found it a powerful organizing principle for its task.

Secondly, the committee commissioned several papers pertinent to its task. These papers examined how other organizations had approached the challenge of identifying future directions for science and technology and what trends they found in science and technology, in the economy, and in the organization and management of industrial research and development. The papers are appended to this report.

Finally, the committee called on multiple resources in making its judgments. It took care to assure that its own membership provided a broad range of expertise and experience. (A list of members of the committee with brief biographies is Appendix A to this report.) And it convened a workshop over 3 days (July 20-22, 2001) at the Science Museum of Minnesota in St. Paul, at which 21 distinguished individuals examined the issues in terms of the "push," "pull," and "contextual" factor taxonomy. The committee is enormously grateful to these participants, who gave up a summer weekend to assist the committee in its task. The workshop agenda, a list of workshop participants, and a summary of the workshop proceedings can be found in Appendixes B, C, and D. In addition to holding discussions at the workshop itself, the committee met three times during the course of the project: May 9-10, 2001, in Washington, D.C.; June 26, in Palo Alto, California; and August 8-9, again in Palo Alto.

Although every attempt was made to ensure a full range of expertise on the committee and at the workshop, the range of potential topics was vast. Some of the differences in emphasis in the report—for example, the number of topics in biological science and engineering compared to those in information science and technology—are in part a result of the kinds of knowledge and experience possessed by the 10 members of the committee and the 21 additional participants in the workshop.

Many people eased the committee's work, and it is difficult to acknowledge them all. However, special thanks go to Laurie Haller of the Science Museum of Minnesota, who in countless and essential ways enabled a successful workshop; to Marsha Riebe of the Hubert H. Humphrey Institute of Public Affairs of the University of Minnesota; and to Maria Jones of the National Research Council, who handled with patience and good humor countless logistical and organizational details for the work of the committee. The committee is also grateful to Michael Casassa and Paul Doremus of the NIST Program Office for their helpful coordination of the committee's work with NIST senior management. Finally, the committee wishes to thank Karen Brown, the deputy director of NIST, for setting before the National Research Council a challenging, at times provocative, and always interesting task.

> Kenneth H. Keller, *Chair*
> Committee on Future Environments for the
> National Institute of Standards and Technology

Acknowledgement of Reviewers

This report has been reviewed in draft form by individuals chosen for their diverse perspectives and technical expertise, in accordance with procedures approved by the National Research Council's (NRC's) Report Review Committee. The purpose of this independent review is to provide candid and critical comments that will assist the institution in making its published report as sound as possible and to ensure that the report meets institutional standards for objectivity, evidence, and responsiveness to the study charge. The review comments and draft manuscript remain confidential to protect the integrity of the deliberative process. We wish to thank the following individuals for their review of this report:

Deborah Boehm-Davis, George Mason University,
George Bugliarello, Polytechnic University,
Matthew Ganz, Navigator Technologies Ventures,
John Hopcroft, Cornell University,
Robert Langer, Massachusetts Institute of Technology,
David J. Lipman, National Institutes of Health,
W. James Nelson, Stanford University,
Steven Popper, RAND, and
Richard N. Zare, Stanford University.

Although the reviewers listed above provided many constructive comments and suggestions, they were not asked to endorse the conclusions or recommendations, nor did they see the final draft of the report before its release. The review of this report was overseen by Robert J. Spinrad, XEROX Corporation (retired), and Alexander Flax, consultant. Appointed by the National Research Council, they were responsible for making certain that an independent examination of this report was carried out in accordance with institutional procedures and that all review comments were carefully considered. Responsibility for the final content of this report rests entirely with the authoring committee and the institution.

Contents

	EXECUTIVE SUMMARY	1
1	INTRODUCTION	7
2	PUSH FACTORS	9

    Biological Science and Engineering, 10
        Molecular and Cell Biology, 10
        Synthetic-Biologic Interactions, 15
        Medical Devices and Instrumentation, 17
        E-Medicine and Health Care Autonomy, 19
        Genetically Modified Organisms, 20
    Materials Science and Technology, 21
        Nanotechnology, 21
        Microelectromechanical Systems, 24
        Fuel Cells, 25
        Materials for Electromechanical Applications, 26
    Computer and Information Science and Technology, 27
        Fundamental Drivers, 27
        System Issues, 28
        Ergonomic Issues, 28
        New Drivers, 30
        Information Technology and Medicine, 31

3	**CONTEXTUAL FACTORS** Evolution of the U.S. Innovation System, 33 Organization of Research, 36 People, 41 Patterns of Investment, 43 Public Policy Issues, 45	33

4	**PULL FACTORS**
	International Challenges, 50
	    Antiterrorism, 50
	    Globalization, 51
	Biological Science and Engineering, 52
	Computer and Information Science and Technology, 54
	Environmental Issues, 57
	International Security and the Global Economy, 59
		49

5	**CONCLUSIONS**	63

APPENDIXES

A	Biographical Sketches of Committee Members	71
B	Workshop Agenda	75
C	Workshop Participants	77
D	Workshop Summary	80
E	Recent Reports on Future Trends in Science and Technology *Michael McGeary*	100
F	Trends in the Economy and Industrial Strength *Kevin Finneran*	129
G	Innovation's Quickenng Pace: Summary and Extrapolation of Frontiers of Science/ Frontiers of Engineering Papers and Presentations *James Schultz*	143
H	Trends in Science and Technology *Patrick Young*	167
I	Trends in Industrial R&D Management and Organization *Alden S. Bean*	199

Executive Summary

In September 2000, the National Institute of Standards and Technology (NIST) asked the National Research Council to assemble a committee to study the trends and forces in science and technology (S&T), industrial management, the economy, and society that are likely to affect research and development as well as the introduction of technological innovations over the next 5 to 10 years. NIST believed that such a study would provide useful supporting information as it planned future programs to achieve its goals of strengthening the U.S. economy and improving the quality of life for U.S. citizens by working with industry to develop and apply technology, measurements, and standards.

NIST recognized that the environment in which it operates is not static. Advances in research are driving technological changes faster and faster. Technological changes, in turn, are leading to complex economic transformations. At the same time, industrial organization is evolving, affecting the processes by which new S&T gives rise to actual innovations. For example, companies are decentralizing their research laboratories and conducting more research through partnerships and contracts. Companies in some sectors are also becoming global, blurring their national identity. Moreover, an increasing amount of innovation is taking place in sectors and companies that conduct little formal research and development (R&D). Finally, social concerns about the effects of new technologies—for example, the impact of information technologies on privacy and the issues introduced by biotechnology and genetically modified organisms—are increasing.

Of course, the future of S&T and its applications is difficult to predict, and transformative breakthroughs that make the biggest difference are the hardest to anticipate. If this study had been conducted in 1991 instead of 2001, for example, who would have predicted the invention of the World Wide Web and its impact on the development of the Internet? (Who before September 2001 would have predicted the impact of terrorism on the homeland of the United States and on the substantial increase in support for antiterrorism research and technology?) Accordingly, the committee was not asked to predict specific future outcomes or recommend what NIST should do. The report, therefore, presents a range of possible trends and factors in S&T, industry, the economy, and society that NIST should keep in mind in its future planning.

The committee proceeded by holding a 3-day workshop, commissioning review papers on relevant topics, and meeting several times to develop this report. Appendixes B, C, and D contain the workshop agenda, the list of participants, and a summary of the proceedings. The commissioned papers are in Appendixes E through I. The 3-day workshop, which took place from July 20 to July 22, 2001, was attended by S&T leaders from a variety of fields (especially from biological, materials, and computer and information science and engineering) and sectors (industry, universities and other nonprofits, and government).

The workshop and this report were organized around three sets of factors expected to shape future trends in science and technology: "push," "pull," and "contextual" factors.

PUSH FACTORS

Push factors are advances occurring or likely to occur in S&T itself. The workshop and the committee focused on three areas in particular—biological science and engineering, materials science and technology, and computer and information science and technology—because it seems likely that many of the important developments in the next 10 years will come from within or at the intersection of these fields. Each is characterized by an extremely rapid rate of change of knowledge; has obvious and wide utility; and will benefit from advances in the others, so that the potential for synergy among them is particularly great. Within the biological sciences and engineering, the successful characterization of the human genome, combined with new techniques for creating, labeling, and analyzing gene microarrays, is likely to lead to rapid advances in the understanding, diagnosis, and treatment of many genetically related diseases. Importantly, research is likely to extend beyond an investigation of DNA sequences to the physical structure of macromolecules, which will advance our understanding of the dynamics of cellular development control pathways and their abnormalities. Much of this understanding and these new technologies will lead to new approaches to drug design. We can also expect that gene sequencing will continue to extend well beyond the human genome and become a tool for studying and modifying other animal and plant species.

Improved understanding of biomolecule structure, combined with new materials development, is likely to lead to greatly increased activity in various aspects of tissue engineering, including the controlled growth of specific biological tissues and the development of hybrid artificial organs. Information technology, from new sensor development to better chips to faster communication links, will give rise to many new microelectromechanical system applications in biomedicine. It will also make possible new approaches to patient data collection, storage, and analysis, with an expansion in both telemedicine and e-medicine.

In materials science and technology, the exploitation of techniques for creating materials with controlled features at nanoscale dimensions will clearly occupy much research attention, leading to materials with unusual and highly desirable physical properties. A second area in which advances are likely is the creation of materials with specific surface properties for use in such applications as catalysts for fuel cells or high-bandwidth fiber-optic cables. Finally, new nonmetallic electronic materials will be developed at a rapid rate, including ceramic, organic, and hybrid materials.

In computer and information science and technology, it is likely that computational speed and communication bandwidth will continue to improve at least as fast as predicted by Moore's law, limited more by economic considerations than by physics. This may stimulate and, indeed, require greater attention to the software development and human-interface issues that are likely to be the bottlenecks in actually utilizing increasing hardware capabilities. What seems clear is that advances in computer and information science and technology will affect the relations among existing technologies, such as cable, telephony, and wireless communications, expanding the potential of each of them, blurring their differences, and requiring a broad rethinking of how they are used and regulated by society.

CONTEXTUAL FACTORS

Organizational, economic, and legal and regulatory issues also strongly affect the S&T enterprise—the patterns of public and private investment, where research is done and by whom, how effective the educational system is, and in what kinds of settings innovation is most likely to occur. These contextual factors are particularly important in understanding where and how public policy can most effectively influence the pace and direction of S&T.

With respect to the research establishment, the next several years are likely to see a continuation of the trend to downsizing or eliminating central research laboratories in large corporations. Outsourcing of development by large corporations and entrepreneurial activity will lead to both an increasing reliance on research within start-up companies and an increase in the number and kinds of cooperative relationships between universities and industry.

Reliance on universities for basic research will continue, but it will be increasingly necessary for that research to be approached in a multidisciplinary fashion, which will represent a challenge to universities, traditionally organized along disciplinary lines. Moreover, universities will be challenged by a continuing dearth of bright American students interested in pursuing research careers. At the same time, closer relations between industry and universities will be facilitated by the growing entrepreneurial spirit of both students and faculty, which will overcome traditional barriers between the two kinds of institutions.

In addition to these trends, the blurred distinction between research and development, the shorter range goals characteristic of small start-up companies funded with venture capital, and the globalization of research and development will give rise to a number of public policy challenges whose resolution will have an important impact on technological innovation. These include issues of government funding of research and development, with particular concerns about, on the one hand, whether adequate investments will be made in long-term basic research and, on the other hand, whether attempts to distinguish research from development will run counter to the dynamics of innovation. There are also policy questions related to a variety of regulatory issues, from antitrust legislation to medical technology regulation to intellectual property protection to standards development.

PULL FACTORS

Pull factors encourage S&T developments in certain directions and discourage, even proscribe, their development in other directions. They encompass a range of national and individual needs and desires, including social and cultural trends and values, increasing concern about the environment, economic and political pressures arising from both domestic and international circumstances, and issues related to globalization—from competition between developed nations on the one hand to the growing pressure on the other hand to deal with the needs of developing countries and the destabilization of the international system that comes from severe economic disparities.

In the wake of September 11, the attention of the nation and the world on the need to combat terrorism and to deal with new kinds of threats to our security will undoubtedly influence the direction of research and development. Technologies will be encouraged—even demanded—that help us to deter, detect, counter, and/or recover from biological and chemical weapons and to combat the networks that support and use them. There will also be a heightened sensitivity to the dual-use nature of many technologies, and the desire to prevent misuse of these technologies will affect every stage of development and adoption. Technology transfer and globalization are likely to be subject to particular scrutiny.

Concerns about the environment appear likely to continue to spur innovation in energy production, materials development, and environmental monitoring and modeling. However, those same concerns may inhibit applications of genetically modified organisms in the food and agriculture sectors.

Both computer and information science and technology and biological science and engineering will be strongly influenced by pull factors because of their significant impact on social structures and culture and personal values. With respect to information technology, tensions involving such issues as privacy, pornography, and free speech are already evident and are made more difficult by the global nature of the Internet and the cultural and political differences among nation-states. New developments in the biological sciences that make possible genetic alteration, cloning, and stem-cell-initiated organ development raise issues of personal and religious values for many people and lead to strong political pressures to regulate research activities and applications in these areas.

All of this suggests that pull factors will be increasingly important in the next several years in determining the direction of technological innovation. Scientists and government will be called upon more and more to communicate with the public about these issues in order to promote a reasoned and informed dialogue and an orderly decision-making process. Furthermore, there will be an increasing need for the educational system to bring nonscientists to a level of understanding appropriate to their involvement in making these societal choices.

The concluding chapter of the report identifies four overarching themes that emerge from the more focused analyses of push, pull, and contextual factors:

- Although it is not possible to forecast what specific advances will be made or when, progress in science and technology will continue to be extraordinarily robust, offering substantial benefits to society and expanding opportunities for further progress. The report examines many examples of promising research advances and technological developments.
- The amount and direction of research and technology development are shaped by the institutional, social, economic, and political environment or context in important ways, including government investment, tax, and regulatory policies; institutional arrangements; and social values. Some areas in which research and technology advances seem feasible may be limited or proscribed because of concerns about privacy (in the case of information science and technology) or the consequences of genetic manipulation (in the case of biological science and engineering).
- Pull factors driven by national needs and consumer demands also play a large role in shaping science and technology. The United States and the rest of world face a number of problems that science and technology could help resolve or mitigate. Individual consumer preferences and needs also affect the demand for research and technology development.
- Although it is possible to discuss trends in science and technology and the factors that affect them, uncertainty about the future remains very high. Uncertainty is inherent in the nature and timing of research advances and technological

innovations, and a number of contextual and demand factors discussed in the report also affect trends and make it impossible to predict outcomes with any precision. There are ways, however, for those supporting or conducting R&D to develop plans that are adaptive in their design and thus more robust against a range of alternatives. The adaptiveness of the system would also benefit from more coordination among the different institutional sectors of the national innovation system (industry, academia, the nonprofit sector, and government) and from the increased technical literacy of citizens.

1

Introduction

The Committee on Future Environments for the National Institute of Standards and Technology, in responding to its task, was again reminded of the enormous breadth and complexity of the science and technology enterprise. The advances in science and technology that actually manifest themselves as changes in our society—as new products or other innovations, as new capabilities, or as new benefits and challenges—depend on a number of factors. Important among them is what advances are occurring or are likely to occur in science and technology itself—what the committee has called the "push" factors. But there are organizational and dynamic issues as well that strongly affect the research and development enterprise—the patterns of public and private investment, where research is done and by whom, how effective the educational system is, and in what kinds of settings innovation is most likely to occur. These "contextual" factors are particularly important in understanding where and how public policy can most effectively influence the pace and direction of science and technology.

Finally, it is necessary to take account of the "pull" factors, those that encourage S&T developments in certain directions and discourage, even proscribe, their development in other directions. These factors include social and cultural trends and values, growing environmental concern and sensitivity, economic and political pressures arising from both domestic and international circumstances, and issues related to globalization—from competition between developed nations to the growing pressure to deal with the needs of developing countries and the destabilization of the international system that comes about because of severe economic disparities.

"Push" factors, "contextual" factors, and "pull" factors were considered separately in the committee's work and are presented separately in the following sections of this report. Although there clearly are interconnections and overlaps between and among the three categories, the committee believes that this approach allowed the most methodical and comprehensive approach to its task. Ultimately, however, it is necessary to bring the issues back together, to synthesize the findings, in order to come to some useful understanding of how different events and choices in the next decade will influence the science and technology enterprise. Therefore, in the last chapter of the report, "Conclusions," the committee has attempted to link and encompass the many issues raised by identifying a set of overarching themes that it believes bear consideration in planning the continuance in the decade ahead of the remarkable achievements of science and technology that we have seen in the past.

The committee discusses a number of technology-related trends in this report, but forecasting the future is difficult and rarely completely successful. It is possible to identify the current status of a technology and to discuss the directions in which it seems to be headed, given developments in science and engineering research, but what ends up happening will be shaped by factors in addition to the technical possibilities. This complexity is why the committee has organized the report in a way that highlights these other factors—contextual and demand. It has been useful, the committee believes, to conduct this exercise, because doing so helps focus attention on the variables NIST should consider in deciding on policies and designing programs. The committee also identified some cross-cutting themes, which are discussed in the last chapter, and highlighted the need for policy makers, whether in government, industry, or the nonprofit sector, to develop plans that take into account the appropriate level of uncertainty and that can adapt to a range of alternative futures.

2

Push Factors

Progress in science and technology can come from any direction at almost any time, but there is fairly broad agreement that major developments in the next 10 years are most likely to come from within or at the intersection of three broad fields: biological science and engineering; materials science and technology; and computer and information science and technology. Each is characterized by an extremely rapid rate of change of knowledge; each has obvious and wide utility; and each will benefit from advances in the others, so that the potential for synergy among them is particularly great. For example,

- Sequencing the human genome would not have been possible without the enormous improvements in computational capacity in the last few decades.
- Those computational advances would not have been possible without improvements in materials and materials processing techniques.
- What we have learned about interfacial phenomena (physicochemical behavior within a few molecular lengths of the boundary between two phases) in biological systems is contributing to the development of new man-made materials, which, in turn, have allowed us to grow functional biological tissues.

Therefore, in considering "push" factors, the committee has focused on these three fields, identifying the subfields within each that seem particularly ripe for major advances or breakthroughs in the next decade. To ensure that this division into fields does not neglect the strong potential for synergy between the fields, the committee's discussions and the report pay particular and deliberate attention to the ways in which progress in any one field will benefit from progress in the others.

BIOLOGICAL SCIENCE AND ENGINEERING

Advances in molecular and cellular biology lead the list of changes in the biological sciences if for no other reason than that they have opened up new fields. But how the new understanding of molecular and cellular structures and events is used in health care and agriculture, for example, depends upon other advances as well. In health care, changes are afoot in diagnostics, drug design, tissue and organ growth, and artificial organs, particularly those known as hybrid organs. In agriculture, advances in the understanding of nutrition and pest control, as well as increasing concern about the environment, guide strategies for modifying organisms to increase the value of foods and to decrease the environmental insult that accompanies their growth.

The field of biology is progressing at a rapid rate, because the scientific opportunities are many, advances in materials and computer and information science and technologies based on them have enabled exploitation of new opportunities in biological research, and public and private funding has increased at a rapid rate. The following sections discuss some of the important areas of advance in biology, focusing especially on those that are enabled by or interact with contributions from nonbiology fields, or that are methodologies enabling a broad range of biological research, or both. These include macromolecule microarrays (or gene chips), synthetic tissues and hybrid organs, and microsensors. The impact of genomics on the genetic modification of plants and animals is also discussed.

Molecular and Cell Biology

Sequencing of the human genome has been nearly completed, setting the stage for the next set of advances: understanding the role of genes in health and disease and using that knowledge to improve screening, diagnosis, and treatment of disease. Since the fundamental structure of DNA has been known for some time, as have been methods for identifying genes and their chromosomal location, the breakthroughs that allowed, in the brief span of a few years, the sequencing of the human genome and the genomes of other plant and animal species have been largely in the development of experimental and computational methods for extremely rapid data generation and analysis, as well as in the management of enormous banks of data. DNA sequencing rates doubled every 29 months in the mid-1980s, and then every 13 months by 2000.

Sequence data will, however, be only a part of the accelerated flow of information during the next decade, and perhaps not the main part. The new techniques for rapid data generation, storage, and analysis of DNA, proteins, and other molecules and cells are providing the basis for various commercial applications. Entire industries have emerged, perhaps the most notable being the biochip industry, whose diverse technological infrastructure encompasses imaging, materials, and a range of information and computational technologies. This section reviews some of the areas of research that will be contributing to our knowledge of disease and health and to better technologies for diagnosis and treatment.

Macromolecule Microarrays

Macromolecule microarrays, or biochips, are surfaces, usually of a size comparable to a microscope slide, to which an array of extremely small clusters of macromolecules is attached by one of a number of techniques. The small size of each cluster—no more than a microdot—allows thousands of these clusters to be arrayed on a single chip. These chips can then be used for a variety of purposes.

The availability of whole genome sequences has enormously increased the utility of DNA microarrays, and it is now routine to assay simultaneously the expression levels of thousands or tens of thousands of genes on a one-square-centimeter chip and to reperform the assay several times under changed environmental conditions. Such assays (also called expression profiles), each involving tens of thousands of genes, can be performed in a day or less, and they can be automated and run in parallel, allowing the determination in just a few days of expression profiles of dozens of cell types exposed to potentially hundreds of different ligands (drugs, toxins, and so on). Since it may be that a relatively small subset of genes (perhaps as few as several hundred) is of critical importance for understanding the origin of cellular changes induced by external or internal signals, this expression profiling is likely to be used much more frequently in clinical situations than complete genotyping. As a consequence, there is likely to be a much greater focus in the next 10 years on this aspect of genomics—and its related commercial and clinical activity—than on complete genotyping.

Whether or not the number of genes causing major diseases is in the hundreds or thousands, studying gene function and complex cellular behaviors (such as cell cycle transitions, serum effects, and heat shock) in the context of the whole cell or organism may be quite complex and call for large microarrays. It is likely that the way cells interpret extracellular signals is more complex than targeting a single gene and involves hierarchical patterns of protein-protein interactions. For example, defining a ligand for a receptor would be very useful, but transmission of the signal may require specific regulatory factors (such as G proteins) to specify an appropriate biological response.[1] The question is whether biochip development will enable such analyses.

In addition to DNA microarrays, other kinds of biochips are being developed to serve as miniature chemical laboratories—"labs on a chip"—having a characteristic feature size of tens of microns. While DNA microarrays are used primarily to monitor changes in patterns of gene expression, a lab on a chip can be used

---

[1]G proteins are in receptors on the surface of cells. Their reaction with a hormone or neurotransmitter is a signal that triggers certain intracellular processes, such as gene expression or changes in metabolic pathways.

to fractionate, mix, or otherwise manipulate microliter quantities of chemicals or small populations of cells. In addition, versatile chip technologies are being developed that can be used to assay for a wide variety of ligands, but they typically involve hundreds rather than thousands of targets.

There is no question that the field is growing extremely rapidly. There are some 200 companies with total gross sales that grew from $272 million in 1998 to $531 million in 2000, and that figure is projected to increase to more than $3.3 billion by 2004.

New Methods of Detection in Microarrays

New methods of detection will play an essential role in expanding the applications of microarrays. Quantum dots are a very promising way of identifying molecules because of their special optical properties. These semiconductor crystals of a few hundred atoms emit a single wavelength, the color of the light depending on the crystal size. In principle, hundreds of millions of combinations of colors and intensities can be achieved. In the near future it should be possible to follow the concentrations of large numbers of molecules simultaneously, because each kind of molecule would, when bound with a quantum dot, cause that dot to give off a distinctive wavelength and thus "label" its presence.

Biosensors are another labeling approach that make it possible to dissect protein-protein interactions in cells that are important in understanding signal transduction, drug delivery, and other processes. Biosensors are constructed by labeling two potentially interacting proteins with reagents that produce a signal when they bind. A common approach currently is to use fluorescence resonance energy transfer, in which different fluorescent compounds are attached covalently to two proteins of interest. Individually they each produce a distinctive color emission, but if they bind, the color emission is different.

Methods of detection that do not involve labeling can also be developed for high-throughput, quantitative monitoring, including surface plasmon resonance and a variant technology based on evanescent waves. The former measures the shift in dielectric constant when molecular targets bind to a surface coated with molecular probes; the latter measures the shift in phase between magnetic and electrical components of a single-mode guided wave as it interacts with probe-target complexes on the surface of the guide. A somewhat more advanced technology is based on optical density changes when a material binds to a surface. Finally, new mass spectrometry technologies, such as matrix-assisted laser desorption ionization mass spectrometry and electrospray ionization, and other non-label-based technologies offer promising high-throughput methods for detecting genetic disease predisposition. Such methods may also help to overcome one of the barriers to the implementation of high-throughput proteomic assays (discussed below).

Medical Impact of Microarrays

The hope, reinforced by recent clinical studies, is that microarrays will be useful for diagnosis and disease stratification. As noted above, it appears that this approach will require not comprehensive genomic characterization but, rather, more limited expression profiles. Normal and diseased cells can show differences in the expression levels of hundreds of genes, indicating that the effects of a disease process can be very complex. However, relatively few genes might be involved in disease etiology. The committee expects that over the next 5 or 10 years, perhaps a few dozen chip diagnostics will be focused on, containing relatively small numbers of genes that are associated with diseases of major importance.

On the other hand, since proteins are directly involved in cell function—unlike genes, which are more indirectly involved—proteomic technologies will need to be developed to obtain a reliable understanding of function, including pathways and network topologies. For certain applications, such as drug targeting (see below), proteomic technologies may eclipse genomic technologies; for other applications, such as identifying populations at risk for major diseases, genomic technologies are likely to remain important.

Many have suggested that massively parallel assays for proteins will follow quickly from our success with gene analysis. However, the extension of the method to protein analysis is beset with complications. To start with, the set of expressed proteins is several times larger than the number of genes (because of alternative splice variants). Furthermore, proteins can undergo a number of posttranslational modifications. In addition, there is an insufficient number of easily produced probes, a greater difficulty (than with DNA) in finding substrates to which they can be attached without denaturing, and problems in achieving sufficiently rapid labeling techniques. Therefore, it is likely that attention will be focused on understanding a limited number of proteins, or polypeptide sequences, that are critically important to disease states. An effort has been made to develop mass antibody approaches to screening proteins expressed in cells, but it is difficult to achieve the immunogen purity on which the specificity of the antibodies depends. Mass spectroscopy is an alternative approach that is less expensive than micro-array assays. For example, mass spectroscopy methods for analyzing tissues are becoming available. The amount of information produced by this approach is tremendous, however, and making sense of it will require concomitant advances in bioinformatics capability.

Both genomic and proteomic assays are likely to have their main diagnostic impact on diseases with well-defined phenotypes that are detectable in cells that are readily accessible. In the near future at least, this is likely to constrain applications to most central nervous system associated diseases, including widespread psychiatric illnesses such as schizophrenia and bipolar disorders. The present best hope for precise phenotypic characterization is functional MRI with a temporal resolution of about 100 milliseconds and a spatial resolution of 1 to 2 millimeters. Continued improvement in the ability to map at high resolution areas associated with language, motor function, memory, and audition and in the ability to detect transient events in small regions will gradually enable the enlargement of recognized disease phenotypes, thus providing better stratified populations for understanding genetic correlates. The correlation of phenotype and genotype, in general, is a major challenge for clinical medicine in this age of genetic information.

Beyond diagnosis and testing, drug target identification is an important application. The challenge is to make the connection between gene expression levels and pathways, pathways and networks, and networks and cell function. For example, an important problem is to find the smallest number of genes needed to convert an abnormal expression pattern into a normal one. This information is obviously also valuable for identifying a disease's causative agents. Absent a more complete understanding of network topology, this drug target identification cannot be attacked rationally. Progress on network mapping is rapid and will be accelerated as new mathematical tools for inference are developed and applied. However, some companies' research in this area has been focusing on "toxicogenomics," that is, finding those patients with a genetic likelihood of having adverse reactions to the drug, thus potentially limiting the liability exposure of the companies. It remains to be seen how strongly market forces or other stimuli will encourage companies to put more effort into using network mapping techniques to improve drug targeting beyond avoiding adverse reactions.

Efforts to map inheritable gene sequence variations, called single nucleotide polymorphisms (SNPs), are under way. SNPs vary from individual to individual, and the SNP Consortium, a large public-private venture, and other investigators expect SNPs to act as markers of genes involved in disease and thus permit cost-effective genotyping. A large library of SNPs might enable researchers to understand complex common diseases such as diabetes, asthma, and cardiovascular diseases that involve multiple genes and environmental factors, identify the most appropriate drug therapies, and even predict individual responses to alternative therapies.

Structural Genomics

Communication within a cell relies on molecular recognition—intermolecular interactions. An ability to manipulate interactions by design requires knowledge of the three-dimensional structure of the molecules. Recent estimates indicate that the number of protein domain folds (generic architectures, as opposed to detailed geometric structures) is relatively small—there are fewer than 1300 water-soluble domains in the biological universe. Of these, we currently know approximately 500. It should be possible in the next decade, with an intelligent selection of sequences for structure determination, to find representatives of every fold using crystallography and other methods. If this is done, it will have a pronounced effect on the biotechnology industry and benefits to human health. Our increased understanding of drug targets from genomics and proteomics will make structural analysis of intracellular molecules an even more powerful tool for rational drug design. Understanding protein structure will also aid in the development of biosensors (discussed above) and of peptide mimetics, in which small peptides designed to mimic protein-protein interfaces can be used to probe protein interactions in cells and define their functions.

Synthetic-Biologic Interactions

Understanding the interaction between biological materials—cell surfaces and macromolecules—and synthetic materials is key to a number of important advances. The ability of structural or functional materials to contact blood or other tissue without damaging it is critical for artificial and hybrid organ design. The effect of synthetic material composition, form, and surface properties on normal tissue growth is key to generating replacement tissues or guiding tissue growth in situ. Finding substrates that can adsorb macromolecules without denaturing them is a requirement for biochip development. New synthetic material matrices can create opportunities for controlled drug release systems or selective membrane barriers.

Because of enormous progress in materials science, discussed in detail below, new methods are being rapidly developed for designing materials with well-defined bulk and surface properties at size scales from the molecular to the macroscopic. Instrumentation for analyzing surface properties is providing increasingly detailed information on the dynamics of the interactions between biologics and surfaces, which can be fed back into the materials design process. Some of the areas likely to benefit from these developments are described in the next few sections.

Assays

Surfaces are critical to microarray technologies. For example, arrays manufactured by spotting nucleic acid probes must be prepared so as to anchor the probes while not adsorbing the targets; arrays of protein probes must bind tightly, but not so tightly as to alter the secondary or tertiary structure of a probe. Proteins are generally more sensitive to the properties of substrates than are nucleic acids, and they tend to denature at interfaces, a problem whose severity increases as the ratio of surface area to sample volume increases. At present, different systems use different surfaces, which makes quantitative comparison of assays difficult. The accuracy of gene identification, quantitation of gene expression level, and sensitivity of the various assay systems are not yet uniform across biochip sources. However, finding the best surface for a specific application is quickly moving from art to science. Recent progress in the development of novel alternatives to glass and silicon promises that widely applicable substrates will be

Tissues

The development of methods to grow various tissues is proceeding at a moderate pace. Sophisticated scaffoldings have been developed on which to attach and grow cells. These scaffoldings, frequently made of biodegradable polymers, have very specific geometric forms and are usually impregnated with growth factors or other solutes not only to ensure cell growth but also to control cell orientation and induce patterns of aggregation that mimic natural tissues and promote normal cell function. Epithelial tissue grown this way is already available as a commercial product for the treatment of severe burns, and a great deal of progress has been made in promoting the re-endothelialization of synthetic vascular grafts. There has also been a good deal of laboratory success in the generation of neurons.

These successes have, in almost all instances, involved culturing already differentiated cells. The current experiments with adult stem cells and the recently altered federal policy on the support of embryonic stem cell research promise very rapid progress in the near future in the generation of an increasingly broad range of tissue types, with improvements in both function and longevity of the tissues produced. Progress is likely to depend on our ability to predict the long-term behavior of these tissues from short-term observations of the physical and chemical interactions between cells and the matrices on which they are grown.

Hybrid Organs

Much of the new understanding generated by advances in tissue engineering is leading to progress in the field of hybrid organs, artificial organs created by housing natural cells in devices that allow them to carry out their natural functions while protecting them from the body's rejection mechanisms. Usually this involves a perm-selective boundary through which chemical signals from the bloodstream can pass, stimulating the cells to produce proteins or other molecules that can pass back into the bloodstream but not allowing the active constituents of the immune system to reach the cells. The technical challenges are the design of the perm-selective membranes, the matrix for the support of the cells, and the system for maintaining them in a viable state.

The last decade has seen major advances in designing artificial pancreases that can carry out at least some functions of the liver. New materials and new understanding of mammalian cell and cell membrane phenomena should accelerate progress in this area in the next decade, and stem cell research may well provide a new, much larger source of cell material for these devices, removing what has been a significant limitation in their design thus far.

Transplantation Immunology

It is likely that the problem of permanent tolerance to transplanted tissue will eventually be mitigated. Advances in immunology and molecular biology lead us to expect substantial progress in this direction during the next decade. In particular, progress in understanding anergy—the process by which cytotoxic cells can be made tolerant to, rather than activated by, specific antigenic signals—offers, for the first time, a realistic hope for effective immunosuppression without continuous drug therapy. In addition, progress in pathway mapping, especially as related to apoptosis, or cell death, opens the possibility of targeting specific cytotoxic T cells (cells that mediate rejection) for programmed death.

Even if the rejection problem is solved, however, the limited availability of human tissue or organs for transplant remains a problem. It is a problem more social than technical, in that organ and tissue donation in the United States (and most other countries) is well below the theoretical level. Therefore, if a significant breakthrough in the number of transplants is to occur in the next decade, there will have to be both a solution to the rejection problem and breakthroughs in tissue engineering that allow the synthesis of a large number of human tissues and organs.

An alternative approach is xenotransplantation, the use of animal cells and organs to replace or assist failing organs in humans. This area is heavily funded by the pharmaceutical industry. However, it is also controversial. Objections come from some medical scientists who fear viral host hopping, from bioethicists who raise conceptual objections to actions that "threaten human identity," and from animal rights activists, who object to any exploitation of animals. This last objection, or at least the broad public resonance with it, is somewhat mitigated if primates are not the source of the transplants. However, it seems at best uncertain, and at worst unlikely, that the broad range of objections coming from different directions will be overcome in the next decade.

Medical Devices and Instrumentation

The advances in materials science and information technology that are making such a profound difference in molecular and cellular biology and tissue engineering have been equally important for the development of new, experimental measurement techniques and new medical devices. When combined with new sensors that take advantage of nuclear and atomic signals—for example, nuclear magnetic resonance and positron emission tomography—they allow the imaging of chemical interactions and processes, such as inflammation and substrate metabolism. Indeed, recent initiatives of the National Institutes of Health and the National Science Foundation, as well as private foundations, have already produced molecular imaging data that could allow significant insights into active physiologic and pathologic processes at the organ and cell levels.

These capabilities are being rapidly expanded by the development of microelectromechanical systems (MEMS),[2] which have made it possible to construct sensors that can be implanted as part of therapeutic or diagnostic devices. Such sensors can also travel within the gastrointestinal or vascular systems, reporting image and/or physiologic data about their environment.

The miniaturization of optical devices and electromechanical systems has also made possible new devices for minimally invasive surgery and for remote tissue manipulation. These are already routinely used, but there is every indication that there will be an enormous expansion in their use in the next decade, with microchips embedded in MEMS systems to create quasi-autonomous microsurgical tools. Since these technical advances are also likely to result in reduced morbidity and reduced costs for hospitalization, this is one area of medical technology where the market signals are expected to foster further technical developments.

One implication of these developments is that the distinction between diagnostic instrument, or monitor, and therapeutic device is rapidly becoming blurred. Implanted defibrillators combine the function of heart monitor and heart regulator. Deep brain electrical stimulators, now used for the treatment of Parkinson's disease, can record and analyze signals to determine when they are positioned optimally and then deliver the therapeutic pulses. Microsurgical tools for cataract removal can navigate themselves into position by measuring local tissue characteristics and then perform their surgical function.

The next 10 years are likely to see a great proliferation of such devices. Since, in effect, they reduce the intermediation of the physician in the therapy, they introduce an interesting set of questions about how much information and what set of instructions must be programmed into a device to ensure that the "clinical decisions" the device makes match its "technical skill."

At the other end of the spectrum, the introduction of electronically mediated connections between the physician and the patient—using various real-time imaging techniques or using imaging and electromechanical devices to help a physician in guiding a scalpel, a laser, or a catheter—means that the physical distance between physician and patient may become immaterial for many purposes. This has led to the introduction of telemedicine: diagnosis of a patient by a physician hundreds of miles away using images transmitted either by digital or analog signal and—even—robot-mediated remote surgery.

In this area, as well as in therapeutics, investigators sometimes work for or start companies to develop the devices they are testing, creating potential conflicts of interest. The ethical implications of this and other commercially relevant medical research will have to be addressed to protect the interests of patients, researchers, and research institutions and to avoid patient injuries and public backlash.

---

[2]For further information about MEMS, see page 24.

E-Medicine and Health Care Autonomy

It is not necessary that the remote signal communication involve a physician. A number of companies are working on ways to automate the process by allowing monitors that are implanted in the patient or merely connected to implanted sensors to transmit their signals remotely to computers that can record the information, automatically make adjustments through chips embedded in therapeutic devices implanted in the patient, or advise the patient on changes in drug regimens or on other adjustments that he or she should make. At present, the technology is at an early stage, and there are a number of different approaches to its design and use. What the approaches have in common is that they provide the patient with a certain degree of health care autonomy, which is becoming attractive to more and more patients.

From a technical point of view, one of the greatest challenges in improving the usefulness of these systems is the lack of compatibility among different monitoring systems and the lack of uniformity between these products and large hospital or laboratory information systems. Groups are working on the development of uniform digital data and patient record systems which, if adopted, are likely to stimulate a rapid expansion in the use of these technologies, many with new chip-based diagnostic devices that put the patient more directly in charge of his or her health.

A more immediate force for autonomy is the Internet and the information it provides. Indeed, it has been noted that the most popular sites on the Internet are those that provide health information. Individuals are able to know or to believe they know more about their condition and their options than was previously the case. The Internet changes the nature of the communication between patients and their healthcare providers. Now, patients are much more likely to arrive at a physician's office armed with information about what they believe to be their problem and how they would like to have it treated.

The main concern here is the reliability of information from Internet sites. In addition to the authentic scientific data and professional opinions that are available, much misinformation and commercial information also find their way to patients. If adequate quality control and standards are instituted, the Web will be a major force for autonomy. Two interesting and important questions face the field in the next decade. Who will take responsibility for validating the accuracy of information on the Web? How will that validation system be implemented? Much of the information is based on federally funded research, and federal agencies have Web sites with authoritative and up-to-date health information.[3] But currently there is no mechanism for certifying the contents of nongovernmental Web sites.

---

[3]MEDLINEplus, the health information Web site of the National Institutes of Health, "has extensive information from the National Institutes of Health and other trusted sources on about 500 diseases and conditions" (http://www.nlm.nih.gov/medlineplus/aboutmedlineplus.html).

Genetically Modified Organisms

Progress in understanding the molecular and cellular biology of plant and animal species has paralleled that in humans. Using many of the same high-throughput techniques and data management systems, the genomes of a number of plant species have already been sequenced.

Some of the uses of the newly available genetic information also parallel the way the information is used in human medicine—determining genetic proclivity for certain diseases, understanding the paths of action of certain diseases, and identifying the patterns of gene expression during development. But applications have gone much further in plants and animals because genetic modification has always been a major activity in agriculture to improve flavor, yield, shelf time, pest resistance, and other characteristics. Genomics merely adds a new set of tools for making those improvements. A number of transgenic crops and animals have already been produced, and in the United States, a large fraction of the planted crops in corn and wheat is genetically engineered.

There are three quite different ways in which genomics can be and is being used in new species development. First, rapid genetic assays can be used to quickly monitor the effects of standard cross-breeding, cutting down enormously on the time previously required to grow the cross-bred tissue to a sufficient state of development that its characteristics can be ascertained. These assays can also provide much more information on changes in the genome that are not obvious in easily observed plant characteristics. Some argue that this, in itself, will provide such an improvement in breeding that laboratory modification of plant or animal organisms will not be necessary.

The other two applications of genomics involve the creation of genetically modified organisms (GMOs)—organisms that are modified by introducing genes from other species. In the first of these, the modification is equivalent to that produced by traditional cross-breeding techniques—a new organism whose genetic structure combines desirable features from each of the two parent organisms. The major advantage to this approach is the speed and selectivity with which the new cross-bred species can be created with recombinant DNA technologies. In the second of the recombinant DNA approaches, the organism is modified by the addition of a gene not natural to the species, which may allow the plant to produce a pesticide or herbicide, or alter its nutritional value or its taste or attractiveness as a food.

Social reaction against GMOs has been strong, especially in Europe, and the ultimate determinant of how widely GMO technology will be used in the future may well be political. On the other hand, the commercial potential of GMOs, as well as their value in meeting food needs, particularly in the developing world, and in minimizing some of the environmental consequences of excessive fertilizer, pesticide, and herbicide use suggest that there will be a strong drive in the next several years to establish their safety. It is likely that this effort will stimulate much greater efforts in modeling ecological systems and will certainly require the development of low-cost techniques for measuring trace concentrations of various organic substances under field conditions.

MATERIALS SCIENCE AND TECHNOLOGY

Our understanding of the relation between, on the one hand, the structure of materials at scales from the molecular to the microscopic and, on the other, their bulk and surface properties at scales that cover an even broader range has been matched by a growing ability to use that knowledge to synthesize new materials for many useful purposes. As a result, materials science and engineering has emerged as one of the most important enabling fields, making possible some of the advances in medical care discussed above, many of the performance improvements in information technology discussed below, and innovations in energy production, transportation, construction, catalysis, and a host of other areas. It seems very likely that developments in materials science will continue to come at a rapid rate in the next decade and will continue to play a vital role in many other fields of science and technology. Progress might be more rapid if public and private investments in materials R&D matched the rate of increase experienced by investments in the biomedical and computer and information science fields. Discussed below are examples of what the committee believes are some of the most promising trends in this field.

Nanotechnology

Although the term nanotechnology is relatively new, developments that have made possible products with smaller and smaller features have been under way since the concepts of microelectronics were first introduced in the early 1960s. Progressively finer-scaled microelectronic components and microelectromechanical devices have been produced using optical lithography to cut and shape superimposed layers of thin films. However, these optical methods have a resolution limit of about 100 nanometers (nm).

Nanotechnology aims at fabricating structures with features ranging from 100 nm down to 1 nm. Nanostructures are particularly attractive because physical properties do not scale with material dimensions as the nanoscale range is approached. This means that if the structure, including size, shape, and chemistry, can be controlled in this size range, it will be possible to develop materials exhibiting unique biological, optical, electrical, magnetic, and physical properties.

Self-Assembled Materials

To build devices with features in the nanosize range will require that they self-assemble—atom by atom or molecule by molecule—to form structures that follow a prescribed design pattern. This also allows considerably more customization, making it possible to imitate nature. For example, considerable research is being pursued in examining hydrogen-bonding interactions similar to those in DNA as orienting forces. Applications in which these assembling interactions upgrade low-cost engineering polymers to incorporate a functional component are suggested. Self-assembly and controlled three-dimensional architectures will be utilized in molecular electronics, highly specific catalysts, and drug delivery systems.

Quantum Dots

Semiconductor nanocrystals ranging from 1 to 10 nm represent a new class of matter with unique opto-electronic and chemical properties relative to bulk and molecular regimes. Their strongly size-dependent electronic states enable one to tune their fluorescence spectra from the UV to the IR by varying composition and size. In addition, as inorganic materials, their photophysical properties are potentially more robust than those of organic molecules. This property makes them attractive as luminescent tags in biological systems, as emitters in electroluminescent devices, and as refractive index modifiers in polymer composites. The main hurdle to commercializing quantum dots remains an effective method to synthesize them in large volumes at low cost.

Nanoparticles

The properties of nanostructured materials are a function not only of their molecular constituents but also of their size and shape. Managing size and shape on the nanoscale allows the creation of a class of materials called nanoparticles. These materials have a unique combination of physical, chemical, electrical, optical, and magnetic properties. For instance, nanoparticles have much better mechanical properties than bulk solids. Thus, carbon nanotubes, one form of nanoparticle, exhibit a Young's modulus of 1000 gigapascal (Gpa) (compared with 210 GPa for steel), with a critical strain to failure of about 5 percent. This high rigidity and low mass make it possible to fabricate mechanical devices with very high resonant frequencies, which would be very useful in wireless technology applications.

Carbon nanotubes have other useful properties. They exhibit a very high thermal conductivity—in the range of 3000 to 6000 watts per millikelvin. Moreover, the high thermal conductivity is orthotropic, with heat transport primarily along the axis of the tube. Thus, the tubes could be used as "thermal wires," channeling heat flow in a single direction. Finally, because they have a low coefficient of friction and excellent wear characteristics, carbon nanotubes can also serve as microsleeve bearings in MEMS applications. The controlled nature of their fine structure allows the clearance between the shaft and the sleeve to be made much smaller than the particle size of common bearing contaminants, further improving their performance.

Nanoparticles constructed of other materials are also very promising. For instance, they offer interesting possibilities for significant advances in magnetic storage. Magnetic nanoparticles 10 to 20 nm in size have the potential to increase storage density to 1000 gigabits per square inch (Gbit/in^2). And, the thermal, electrical, and thermoelectric properties of nanostructure alloys of bismuth can be tailored to achieve a marked increase in the thermoelectric figure of merit, which would make possible the design of high-efficiency, solid-state energy conversion devices.

Hybrid Structures

Combining nanoparticles with other structural components provides an even wider range of possibilities. For example,

• By using size-graded nanoparticles together with high-pressure isostatic pressing, a marked increase in the density of ceramic materials can be achieved. The reduced porosity leads to enhanced strength, which has already been exploited in zirconia hip prostheses.

• In a very different biological application, a material with the commercial name Hedrocel is produced by vapor deposition of tantalum on a vitreous carbon matrix. Hedrocel is about 80 percent porous, has an average pore size of 550 microns, and is extremely effective in fusing bone in spinal, hip, and knee reconstruction. It also seems likely that the ability to control both porosity and chemical properties will turn out to be very useful in the synthesis of perm-selective membranes in hybrid organs.

• Nanotechnology has the potential to provide better surfaces or substrates in bioassays for chemical/biological reactions and for research on the organization of interconnected cells, such as neurons and the endothelial cells that line the circulatory system.

• The blending of nanoplatelets of montmorillonite—layered silicate clay—with nylon has significantly improved the mechanical properties of the resulting polymer. These clay-filled composites offer better moldability, ductility, and high-temperature properties.

• Combinations of nanoparticles and electrically conducting polymers make possible surface coatings with characteristics that can be varied with an electrical command. Surfaces that convert from transparent to opaque, change color, and heat or cool have been demonstrated. It is possible to incorporate electrical and optical properties on the surface of composites (coatings or adhesives). These surfaces can be imaged at low cost with inkjet technology, creating many possibilities for new products.

The great potential for enormously expanding the range of achievable materials properties through nanotechnology seems likely to bring about a paradigm shift in the development of materials. Further, it is likely that in the next several years it will become possible to produce anisotropic nanoparticles economically. Coded self-assembly of nanostructures, coupled with established methods of optical lithography, should enable the integration of structures and features with different scales, and processes for assembling these particles into controlled arrays will provide further flexibility in designing a number of novel devices, including a variety of microelectromechanical devices.

Microelectromechanical Systems

The past decade has seen considerable progress in designing microelectromechanical (MEMS) devices, helped by fabrication techniques originally developed for the semiconductor industry. Further progress, however, will depend on solving two classes of problems: those associated with control of and communication with the devices and those related to the surface properties of the materials used in the devices.

The control and communications issues are, in essence, questions of internal and external wiring design, which obviously becomes more difficult as the devices become smaller. It appears that a higher level of integration of the sensors, microprocessor, actuators, receivers, and transmitters in these devices may solve the internal wiring problems and that use of wireless technology may ease or eliminate the need for external wiring. For the latter approach to be successful, however, it will be necessary to achieve miniaturization of transmission and receiving components without an excessive reduction in signal power. This suggests that the ability to achieve high power densities in these components will become an important design focus.

It is a general feature of scaling that the surface to mass ratio of any device will increase as the device gets smaller. As a consequence, the properties of the surfaces of MEMS devices take on great importance, either creating opportunities for using the surface in the device's function or creating problems when the surfaces are incompatible with the medium in which they function. For example, in microfluidics applications, altering the surface energy of the very small channels that comprise the device, perhaps by an electric signal, might be sufficient to allow or prevent the flow of a fluid through a particular channel, thus creating a nonmechanical switching system.

On the other hand, a bio-MEMS device has many of the problems of artificial organs and other implanted devices: the need to avoid damaging the blood or tissue with which the surfaces of the device are in contact and the need to maintain the properties and integrity of its materials of construction over the lifetime of the device if it is implanted. There is a large literature on material-biologic interactions that can help guide bio-MEMS designers, but the need to satisfy simultaneously a set of compatibility criteria and a set of structure/function criteria is likely to be one of the important challenges of the next decade with respect to these devices.

Fuel Cells

Fuel cells offer an efficient, nonpolluting method for generating electricity for both mobile and stationary applications. For the past decade a significant effort has been made to commercialize the proton exchange membrane (PEM) fuel cell, which has the advantage of lower costs and better operating conditions than other types of fuel cells that operate at higher temperatures. PEM fuel cells incorporate three primary subsystems: a fuel source that provides hydrogen to the fuel cell, a fuel stack that electrochemically transforms the hydrogen into DC current, and auxiliary systems that condition the output and regulate the cell. All three subsystems will require significant improvements in cost, size, and durability before widespread applications become feasible.

Catalysts—platinum and platinum alloys—used for PEM fuel cells are not only expensive ($20,000/kg), but their long-term durability has not been established. Reduction in the amount of the platinum required and/or the use of a less expensive catalyst would reduce cost significantly. Research on the effect of lower catalyst loadings and on the stability of catalysts finely dispersed on carbon supports is under way, as is research on the development of non-platinum-based catalysts such as organometallic porphyrins and nonprecious metals.

Because this country does not have a hydrogen distribution and storage infrastructure, at present fuel cells are fed by hydrocarbons that are transformed by a fuel processor into hydrogen gas, CO_2, and CO. At temperatures less than 120 °C, the absorption of CO on the platinum catalyst is significant, and it competes with the oxidation of the hydrogen on the fuel cell anode. To reduce the effect of this CO poisoning, research is under way to develop anode catalyst alloys that are less susceptible to CO poisoning and to add CO cleanup stages to the fuel processor. Direct hydrogen-fed fuel cells would eliminate these problems, but the infrastructure for hydrogen distribution and storage is not in place and will require a tremendous investment to establish.

Current membrane materials, such as perfluorinated sulfonic acid, cost about $500/m^2. Because these membranes require a high level of humidification for durability and for proton conduction, the cell stack operating temperature is constrained to 80 to 90 °C. This is a serious constraint. If the operating temperature of the PEM could be increased to 120 to 180 °C, it would be possible to eliminate the CO poisoning problem. Also, at the higher temperatures heat exchangers are more effective and heat recovery for cogeneration applications becomes feasible. However, operation at the higher temperatures would require a membrane material that could conduct protons without the presence of water, a major research challenge in the next several years.

Car manufacturers are aggressively pursuing research to lower the costs of fuel cell systems. Predictions of the availability of fuel-cell-powered automobiles vary widely. Toyota will reportedly introduce a fuel-cell-powered automobile on a limited basis by 2003. The target of U.S. car makers is much later (2010). Currently, they are attempting to reduce the cost of a fuel-cell-powered motor to about $50 per kilowatt (kW), or about $3500 for a 70-kW engine.

Several companies are preparing to market stationary units for homes with prices ranging from $500 to $1500 per kW. If research makes possible the operation of fuel cells (and on-site hydrogen conversion units) at costs of 0.10 per kWhr or lower, then distributed generation of electricity production will become feasible. Even at the current anticipated cost of $0.30 kWhr, the stationary home units are expected to be welcomed by those living in remote areas who are not connected to the electrical grid.

Materials for Electromechanical Applications

As noted earlier, advances in information and communication technologies have always depended strongly on developments in materials science. Better sensors, higher bandwidth communication links, finer chip features for more rapid calculation speeds, and storage devices have all relied on novel and improved materials. There is no reason to expect either the importance of this area or the rate of progress in it to diminish over the next decade. Some of the most promising developmental areas are described briefly below.

Photonic Crystals

One-dimensional photonic crystals (dielectric mirrors, optical filters, fiber grating, and so on) are well-developed technologies used to manage light. In the near future, materials and processes to prepare two- and three-dimensional photonic crystals and methods for integrating these devices on a chip compactly and effectively should be available. These developments will result in improved optical functionality to meet the needs of high-bandwidth communication networks.

Materials with Improved Dielectric Properties

It is likely that more new materials with improved dielectric properties (porous silicon is an example) will be developed and will support the continued miniaturization of integrated circuits. Materials with higher dielectric constants are important for the dielectric layer in capacitors and transistors. Materials with lower dielectric constants will be valuable for insulating the metal interconnects on the chip and on the circuit board more efficiently, decreasing separation requirements and improving transmission speeds.

Organic Electronics

Employing organic materials as active layers in electronic devices will become increasingly important. These materials will allow the fabrication of inexpensive devices using nontraditional semiconductor manufacturing processes. Device drivers, RF-powered circuits, and sensors are all potential applications. Other electronic devices, based on the modulation of the conducting properties of single organic molecules, could significantly advance circuit miniaturization.

Organic/Inorganic Hybrids

The motivation for developing organic/inorganic hybrids is the coupling of the improved processing and superior mechanical properties of the organic material with the unique functional characteristics of the inorganic material to form the hybrid material. Semiconductor applications appear possible in which the charge transport characteristics of the inorganics (because of strong ionic and covalent intermolecular associations) are vastly superior to those of the organics (which generally only have relatively weak van der Waals interactions). Electronic, optical, and magnetic properties that are special or enhanced in organics can also be utilized in the hybrids.

COMPUTER AND INFORMATION SCIENCE AND TECHNOLOGY

Fundamental Drivers

Because a major driving factor in computer technology has been its continued ability to follow Moore's law—that is, an approximate doubling of speed every 18 months, with concomitant increases in capacity and decreases in costs—the committee does not foresee a significant slowing of this trend over the next decade. The economics of computer fabrication (for example, lithography) may be a limiting factor, but it seems unlikely that physics will be a limiting factor in the next decade. Increases in storage density and decreases in cost are also expected to follow the trends of the last few decades, which currently exceed those predicted by Moore's law. It is less clear that the increase in consumer demand will match the ability to produce faster computers embodied in Moore's law, but consumers will still value a broader set of computer and network applications and more reliable and secure software. Computers will also be increasingly embedded in everyday products. Hence the committee expects to see the continued adoption of computer solutions to an increasingly large number of applications, including those made possible by faster calculation speeds and those made economically practical by decreasing calculation costs.

Communication technology using both copper and optical fiber has seen a comparable rate of change. In communications it is expected that the "last-mile" problem will be solved; that is, the bandwidth limitations and administrative barriers to home Internet connection will not impede deployment. (It was observed that cable connections were provided to the most of the nation in the last few decades, so there is no reason to believe that a demand for higher bandwidth connections—or better utilization of cable—cannot be met with already existing hardware. The main question is the extent to which the Internet services available are actually wanted by the public.)

How best to use the limited radio frequency spectrum for mobile computer communications is a challenge for the next decade and will be an important issue for standards as the technology is increasingly used for transnational communication. The problem may be eased or solved in part by technological advances—for example, in protocols and compression—but it is likely to require regulatory action that may, in turn, have a significant technological impact.

System Issues

While the capabilities of computation and communication will continue to increase, the complexity of larger and larger systems may limit the rate at which these increases can be effectively utilized. While incremental advances in current applications can capitalize on the improved technology, the implementation of new applications, potentially possible because of hardware technology improvements, is likely to depend on improvements in software development. Historically, more than 50 percent of the cost of software development has been in testing, debugging, and quality assurance. Reducing those costs will obviously be a high priority and will provide a strong impetus for the further development of the software engineering discipline as well as the standardization of the software development process. It will also be helped by improvements in automated software production as well as improvements in self-diagnosis and self-correction. Many would also argue that, as complexity increases, software standards and software interoperability will assume increasing importance. Also, there will be increasing emphasis on ensuring the security of software as computing becomes increasingly Web-based and thus more vulnerable to malicious hacking and cyberterrorism. Another area of growing emphasis will be the reliability of complex programs used in safety-critical applications, such as air traffic control and nuclear power plant operations.

Ergonomic Issues

The (short) history of widespread software use suggests that human-system interface issues will strongly shape, even limit, the adoption of new applications. Computer languages, operating systems, and applications that were not natural and easy to learn have struggled to gain widespread acceptance and have not always survived the introduction of those that were (although once people have begun using a technology, they are reluctant to switch to a new technology even when it provides better usability). ("Natural" in the context of something as artificial as computer systems is somewhat of a misnomer, but it means that the interface must build on actions already familiar to the user from other applications or systems.) The power of the computer can be used to provide a relatively simple interface to the user in spite of the underlying complexity of the system. Unless there are major advances in the design of these interfaces, commercial demand will lag the technical capacity for increased complexity or further performance breakthroughs in computers and networks.

The cellular telephone is an example of technology that has been adopted fairly rapidly, in part because the average consumer can operate it despite the sophisticated computation and networking involved in the underlying system. In this case, the user finds the interface familiar, because it builds on interfaces that most have seen before—the ordinary telephone and the menus used in graphical user interfaces of computer operating systems, for example. Even with cellular telephones, however, people have problems using the menus, and efforts continue to make them more usable. The spreadsheet is another example of a technology that achieved widespread use in part because of its functionality but also because its operations are fairly familiar to most people. Most first-time users can quickly grasp data entry procedures and how to perform at least simple arithmetic operations and use common formulas.

The problem is that interfaces that build on already-understood technologies (i.e., so-called WIMP interfaces—windows, icons, menus, pointing) to avoid the learning problem inherently pass up opportunities for innovation. It seems likely that new interface modalities will not accept this constraint, even though it is not clear what technical approach they might take and how they will meet the requirement of being user-friendly. Voice input and output might become a factor in some applications, because the technology will soon be capable of natural-sounding speech with computing resources that are modest by present-day standards. Voice will not be the only modality on which to focus. Others include pressure-sensitive input devices, force-feedback mice, devices that allow people to "type" by tracking eye movements, and wearable computer devices. The effect such devices have on performance will be an important issue, especially in safety-critical systems, where an error could have catastrophic consequences.

Innovative application of artificial intelligence technology and advances in sensors that are able to integrate information may greatly improve a system's ability to interpret data in context. For example, speech recognition is improving very rapidly. However, it must be noted that general speech understanding under natural conditions requires a degree of context that will not be possible within at least the coming decade. If speech becomes a major general input modality in that time frame, it will be because a subset of natural language has become an accepted computer interface language.

Visual input will assume more importance in specific contexts, but the problem of general visual recognition will still be difficult even with 100-fold CPU speed advances expected during the next decade. Nevertheless, existing capabilities presently used routinely by the military (for example, for target recognition from moving aircraft) will find some similar specific applications in the civil sector, such as traffic safety applications.

Some of the interface problems, especially contextual speech recognition, may be alleviated if the interface is readily personalized for each user. For applications in a mobile setting (when limited bandwidth is available), this will become possible when all users are always carrying large amounts of personal data—made possible by the increases in storage technology. By greatly improving voice recognition, and providing abundant context, these personal data might permit advances in interfaces that would otherwise be impossible, perhaps also enabling less sophisticated users to tap into high-powered computers and networks.

An important part of the solution to the user interface problem will probably lie in building computing and networking into appliances in ways that are invisible to the user. To the extent this occurs, the need to improve human/computer interfaces is obviated.

Greatly improved technology for data transmission and storage naturally raise issues of security and authentication. The committee believes that the technical issues will be solved (that is, the level of security and authenticity appropriate to the application will be affordable) but that privacy issues are not just technology issues. They involve trade-offs and choices. For example, someone carrying his or her complete medical record may want total control over access to it, but if the person arrives at an emergency room unconscious, he or she would probably want the medical team to have relatively unrestricted access. Thus, it is not a technology problem but a policy issue. Similarly, network devices will make it possible to identify the location of the user, which the user may not want known, at least without giving permission. Technology will be capable of providing the needed encryption. Society must decide what it wants. This is an issue discussed in more detail in a later section.

New Drivers

An important question is whether there are drivers on the horizon other than continuing improvement in the speed of computation, the capacity of storage, and the bandwidth of communication links that will give rise to new kinds of applications. The committee has not identified any but is of the view that the convergence of various communication and computation technologies that commingle traditional broadcasting (cable and wireless), telephony, and Internet venues, coupled with improvements in each, may well lead in time to a significant expansion in certain kinds of applications. For example, one might foresee great improvements in the quality (and therefore the attractiveness) of remote, real-time social interaction. Primitive telepresence can be expected to be available within 10 years—an enhancement of videoconferencing that is likely to make it an increasingly attractive substitute for some face-to-face communication. However, improvements of several orders of magnitude in the speed of computation, communication, and display will be needed before telepresence will be an entirely acceptable substitute.

A number of observers have suggested that the integration of computers with traditional broadcasting media could complete the process of establishing a continuum or seamless web of communication possibilities, from one-to-one contacts (the telephone) to one-to-many contacts (broadcasting) to many-to-many contacts (digital networks). Moreover, with this continuum in place, the potential for a number of new types of group interaction and social group structures (electronic polling, electronic town halls, new modes of remote education, quality program support for interest groups, and so on) increases. Although many of these new interactions may ultimately become realities, it is the committee's belief that, in the near term, entertainment applications such as on-demand movies are most likely to be the major beneficiary of the integrated technologies.

Information Technology and Medicine

One premise in the choice of the three broad fields that the committee considered in describing technological "push" factors was not only that each was a field in which knowledge was expanding rapidly but also that each had widespread impact on the other two and on technologies outside the three fields. That point is made in a number of the examples discussed above. However, the intersection of information technology and biomedicine is of such profound importance that it deserves specific emphasis and even some recapitulation.

Computation has the potential to dramatically speed the acquisition of new knowledge in the biological and medical sciences by making it possible to extract meaning from large data sets. In molecular and cellular biology, the rate at which this happens will depend on further progress in the development and successful marketing of the high-throughput assays and imaging technologies that generate data on systems, rather than data on isolated entities. It is also dependent on continued advances in computer science and technology required to organize the data and disseminate it—that is, to convert it to useful knowledge. A driver of genomic analysis has been the ability to process the enormous amount of information produced and identify important results. This capability for computing power is likely to have to increase exponentially to achieve a similar level of proteomic analysis. New systems for classifying and organizing data and for creating knowledge and knowledge bases, rather than databases, are needed.

Computational power has also changed the pharmaceutical industry, where for some time it has been providing support for three-dimensional modeling of molecular structures as part of the drug design process. It has also unleashed the power of combinatorial chemistry in the search for new drugs, providing, as in the case of molecular biology, the means for ordering and analyzing huge data sets. Another promising direction in the future is the modeling of organ systems and whole organisms, a field that has been labeled physiomics. Here, computational power provides the ability to capture a vast number of parameters related to the physiology of an individual, to model that individual's biological responses to a drug or a combination of drugs, and thereby to customize drug therapies to an individual, reducing the uncertainties and side effects that are a feature of current drug use.

Physiomics is not without its challenges and problems. Since the validity of the models it generates and manipulates depends on the accuracy and completeness of our understanding of the biology response functions involved, the move from a clinical or laboratory setting to a computer terminal must involve a true bridging of the biomedical sciences and information technologies, a partnership between scientists in both disciplines, and the capacity to communicate in both directions.

At the clinical level, information technology has made possible major advances in imaging techniques, and there is every indication that the next decade will see a continuation of that trend. However, it is not so much overwhelming computational power that is most important as it is our ever-increasing ability to miniaturize chips that have modest, but adequate, computational capacity to be usefully incorporated in implanted sensors or in MEMS devices. Intelligent sensors based on microchips are now commonplace in heart pacing and implanted defibrillation devices. Insulin pumps can be driven by implanted glucose sensors. But microchips are likely to provide the key enabling technology in the next several years by allowing the design of MEMS devices for use in microsurgery, motor control in replacement body parts, in situ drug delivery, and a host of other applications.

Finally, there is the enormous potential for information technology to facilitate the storage and transfer of information about patients between and among different departments of a health care facility and between health care facilities. It has already been demonstrated that information technology has the potential to reduce errors in drug dispensing; to automate patient monitoring either directly if the patient is in a health care facility or through a simple analog telephone connection if the patient is at home; to store and transmit x rays and other images from one facility to another; or to store a patient's entire medical record on a wallet card. The fact, however, is that health care has seriously lagged other industries in incorporating information technology into its operations. That, combined with the opportunities that will come in these next years as computational speeds and communication bandwidths increase, suggests to the committee that there will need to be significant catch-up in this area in the next decade.

3

Contextual Factors

Scientific research and technology development do not develop in a vacuum. They are shaped by contextual factors, such as institutional and funding patterns and social values. Although these factors are discussed generally in this section of the report, they would affect the three fields of focus in this report somewhat differently. Each field has a distinctive organizational pattern, a different emphasis on science and technology development, and a different mode of operation. The biomedical field, for example, has a substantial university component and is driven the most by basic research. The computer and information technology field, on the other hand, consists mostly of small firms and is largely technology driven. Materials occupies the middle ground. The work is more likely to be carried on by large and established firms, and there is more of a balance between basic research and hands-on technological effort.

EVOLUTION OF THE U.S. INNOVATION SYSTEM

The R&D enterprise in the United States is carried out as a public-private partnership, with roles for each sector in funding the effort and in carrying out the actual work. The system has evolved since the end of World War II, and not only does the committee expect it to continue to do so, but there are indications that the rate of change will accelerate as a result of significant changes in the environment for R&D. Federal funding of curiosity-driven research is under pressure, changes in corporate structure and the emergence of new industries are giving rise to changes in the organization of industrial research, patterns of private investment in R&D are changing, the globalization of R&D is giving rise to new competitive pressures, universities are becoming more entrepreneurial, and the career choices of undergraduate and graduate students are changing (there is less interest in science and technology careers, and among those who do go into science or engineering, there is more interest in jobs outside academia). In this chapter of the report, the committee addresses what it believes will be among the most salient elements in shaping the environment over the next several years for science and technology in the United States.

A useful starting point is a brief summary of what the system looks like today and how it evolved. The R&D enterprise in the United States had grown to an estimated $265 billion a year in 2000 from a starting level of less than $5 billion a year in the years shortly after World War II.[1] Until 1980, federal funding exceeded industrial funding, but by 2000, industry accounted for 68 percent of R&D expenditures and carried out 75 percent of the work, measured in dollars spent. Most of those expenditures, however, could be categorized as development; basic research comprises only 18 percent of the total spent nationally on R&D and the patterns of funding for and performance of basic research were and continue to be entirely different from those of development.[2]

Federal support of R&D is provided by a number of agencies. Much of the funding, especially for applied research and development, is provided by the mission agencies that are users of the results, for example, the Department of Defense and the National Aeronautics and Space Administration. Some, including the National Science Foundation and the National Institutes of Health, conduct or support research for its own sake, including basic long-range research, although funding is generally predicated on the historical contributions of research to national well-being. Some R&D agencies exist to support the private sector, such as the Cooperative State Research, Education, and Extension Service in the Department of Agriculture and NIST in the Department of Commerce. NIST not only supports technology development, it is also responsible for metrology, the science of weights and measures, which underlies the development of technical standards relied on by industry.

In 2000, the federal government provided 49 percent of the funding for basic research, industry contributed 34 percent, and another 18 percent was provided by the universities, other nonprofits, and nonfederal governments. Although the federal government was the largest funder of basic research, it was not the largest performer. Universities and university-based federally funded R&D centers (FFRDCs) carried out 49 percent of all basic research in the United States in

---

[1]Research and development data in this and the following paragraph are from National Science Foundation, *National Patterns of R&D Resources: 2000 Data Update*. NSF 01-309. Arlington, Va.: National Science Foundation, 2001.

[2]The categorization of R&D expenditures into basic research, applied research, and development has a certain element of arbitrariness and has been the subject of much discussion. Therefore, these figures should be treated as no more than approximate.

2000, industry and industry FFRDCs 34 percent, nonprofit research institutions and nonprofit FFRDCs 10 percent, and federal laboratories about 7 percent.[3]

Basic research activity is far from uniform across the more than 3000 postsecondary institutions in the United States. Most university basic research is carried out at just 60 universities, both public and private.[4] Industry provides 8 percent of funding for academic basic research, a level that changed very little in the last decade. In contrast, the federal government provided about 58 percent of the funding for academic basic research during the last 10 years (the rest is funded by the universities themselves and other nonprofits). Thus, research universities continue to rely very heavily on the federal government.

There have been clear trends in recent years in the research fields supported by the federal government.[5] In the mid-1990s, expenditures on nondefense R&D moved ahead of defense expenditures for the first time. That trend has continued, at least until now. Spending on health dominates, accounting for almost half of the federal nondefense research budget. Health research has also shown by far the fastest growth, compared with other fields of research, which as a group have had nearly flat appropriations in recent years, as measured in constant dollars. Nonhealth research increased by less than 0.6 percent from 1993 to 1999, while health research increased by 28 percent, in constant dollars. Federal funding of math and computer sciences increased by 45 percent, but funding of the physical sciences was 18 percent less in 1999 than in 1993. Several fields of engineering also received less. A recent study of these trends by the Board on Science, Technology, and Economic Policy (STEP) of the National Research Council (from which the data in this paragraph were taken) concluded that there is cause for concern about reduced federal funding of most of the physical sciences and engineering, because it is being driven by changes in the mission of individual departments, especially the Department of Defense, not from a systematic analysis of national needs and priorities. STEP recommended that the White House Office of Science and Technology and the Office of Management and Budget lead an

---

[3]NSF, *National Patterns of R&D Resources: 2000 Data Update,* Tables 2A, 2B, NSF 01-309, at <http://www.nsf.gov/sbe/srs/nsf01309/start.htm>.

[4]The survey of academic R&D does not break out basic research. This statement is based on the fact that in 2000, 60 universities accounted for nearly two-thirds of the funding for academic R&D and the assumption that basic research funding is probably even more concentrated than funding for academic applied research and development (NSF, *Academic Research and Development Expenditures: Fiscal Year 2000* [Early Release Tables], NSF 02-402, Table B-32, at <http://www.nsf.gov/sbe/srs/srs02402/start.htm#rd4>).

[5]An analysis of shifting emphasis in the federal funding of major fields of research and the implications is provided in National Research Council, Board on Science, Technology, and Economic Policy, *Trends in Federal Support of Research and Graduate Education.* Washington, D.C.: National Academy Press, 2001.

evaluation of the federal research portfolio to ensure that research related to industrial performance and other national priorities is adequately supported.[6]

Finally, it should be noted that the past couple of decades have seen a sea change in industrial laboratories, which, as noted below, appears likely to continue. Many of the great industrial central laboratories—Bell Labs, GE, Exxon, to name just a few—have been significantly downsized, or have reoriented their research to shorter-term goals, or both. There have also been changes in university laboratories, or at least in the strategies of universities, following the passage of the Bayh-Dole Act in 1980, which allowed universities to acquire intellectual property rights on products and processes even if they were developed with government funding.

Given the present state of affairs and the trends evident at the moment, the question is what the effect will be on research and development activities in the next 10 years. The following sections address that question.

ORGANIZATION OF RESEARCH

To comprehend the likely changes in the organization of research in the next decade, it is necessary to sort through the trends in the levels of public and private support for R&D as well as the kinds of R&D that public and private entities are likely to provide. As corporate research centers continue to decline because of mergers or pressure to maximize short-term profits, even more basic research will have to shift to universities. Since there is nothing in recent trends to indicate that industry will increase its support for university research to a substantial amount (despite a few well-publicized individual university-industry linkages), it is likely that the government will remain the principal source of funds.[7]

There are some serious challenges and uncertainties in this respect. Although federal funding for research has been growing, the growth has been far from uniform across the various fields of science. Biomedical science support has grown substantially, which will probably continue, but other areas have not fared as well. Since the thesis of this committee is that exploiting synergies between fields is especially important in achieving scientific progress, success in the next decade is likely to depend on either redressing the imbalances in funding or changing the approach to research support within specific agencies. In regard to the latter option, the National Institutes of Health (NIH) is particularly important. As was pointed out in Chapter 2, there are enormous opportunities to advance the

---

[6]National Research Council, Board on Science, Technology, and Economic Policy, *Trends in Federal Support of Research and Graduate Education*. Washington, D.C.: National Academy Press, 2001.

[7]Industry provided 6.8 percent of university R&D funding in the early 1990s and about 7.2 percent in the late 1990s (National Science Foundation, *Academic Research and Development Expenditures: Fiscal Year 2000* [Early Release Tables], NSF 02-702. Arlington, Va.: NSF, 2001).

biomedical sciences, but both materials science and information technology are key to many of those advances. Although NIH is clearly the main source of federal support for research in the biomedical sciences (88 percent in 1999),[8] the organization of its institutes around specific diseases and its general expectation that research questions will be structured around specific hypotheses about the cure of those diseases make it difficult to pursue the kind of multidisciplinary research described earlier, which tends to be structured around questions of physicochemical mechanisms, new approaches to virtual experimentation, or device and instrument development. The institute structure also fragments research on cross-cutting problems such as medical error, nutrition, and risky behavior.

A second source of uncertainty about how the federal research support system will function in the next decade is the sporadic but continuing debate in Congress over the value of establishing explicit goals for research or expected outcomes from it. For the past several years, the idea that the best guarantee of progress in science and technology is to support curiosity-driven research has been questioned by a number of people in Congress, including some of the strongest supporters of science. The argument is that the high cost of research requires that investments be guided by clear goals, which may range from pursuing areas deemed ripe for progress to identifying social needs and encouraging focused efforts to meet those needs, and progress toward those goals should be assessed on a regular basis.

In fact, much federally supported research has always been in specific areas of national need, notably national defense and health. Since the end of the Cold War, at least until September 11, there has been more emphasis on R&D to support national economic innovation, in which—superficially, at least—the notion that progress can be measured seems possible. It seems likely, therefore, that an increasing fraction of research, particularly when it is multidisciplinary, will at least be described in terms of and possibly subsumed in goal-oriented projects. At the same time, history shows that important advances often come from discovery- rather than goal-oriented research. Given the uncertainty of research, a healthy research portfolio will be diversified among mechanisms as well as among fields. In this regard, the National Nanoscale Initiative of the National Science Foundation (NSF) may be a harbinger of things to come. One benefit of a goal-oriented approach is that it tends to encourage multidisciplinary efforts, which have always faced difficulties in the academic world, but it will be a challenge to ensure that high-risk, long-term, or open-ended research efforts are not seriously impeded.

The interest of universities in creating intellectual property and the growing attractiveness of entrepreneurship with its financial rewards for both faculty and

---

[8]NSF, *Federal Funds for Research and Development: Fiscal Years 1999, 2000, and 2001*, NSF 01-328. Arlington, Va.: NSF, 2001, Table C-34.

graduate students are also likely to continue. Although these are not entirely antithetical to long-term basic research—for example, the development of new instruments or new synthesis methods often occurs in the context of an ongoing research program with more fundamental goals—there is likely to be some shift in university research from long-term basic research programs to programs focused on short-term results.

Although there are some downsides to this shift, it is also likely to bolster cooperative efforts between U.S. universities and industry, creating more opportunities for technology transfer and cross-fertilization, mirroring relationships that have heretofore been more prevalent in Europe and Japan. We can certainly expect to see a continuing increase in industrial licensing of university-developed intellectual property. Industry will look to universities for students trained in the skills sets they need, and it will have professors serve as consultants or actually work on research projects. More professors will take sabbaticals to work in industry, and an increasing proportion of graduate students will spend time as interns.

The intensified interactions will require rules and guidelines related to conflict of interest and ownership of intellectual property, but there are already signs that the growing relationships between U.S. universities and industry are helping all parties to sort through those issues and to develop acceptable approaches. For example, it is now commonplace for research universities to have formal and comprehensive policies on conflict of interest, along with standing committees to monitor compliance. Although it would be premature to say that all of these issues have been resolved, it appears to this committee that the learning curve will continue and, although concerns will not entirely disappear, relatively standardized patterns and protocols for these university-industry relationships will soon be in place.

We can expect that government agencies and programs will promote and support cooperative efforts involving a number of corporations and universities in high-risk, high-reward research with specific goals. These programs will often encourage or even require multidepartmental or multidisciplinary collaborations within a university. A typical example is the Multidisciplinary Research Program of the Department of Defense's University Research Initiative. The program specifies areas of interest (for FY 2002, for example, it lists 19 research areas, including integrated nanosensors, membranes based on active molecular transport mechanisms, fuel cells, and the behavior of scaled-up information networks), requires the involvement of researchers from more than one discipline, encourages multiuniversity collaboration, and makes clear that successful applications will have to either involve joint university-industry research or provide in some other way for transferring technology to the private sector.

Research will not only be more multidisciplinary, it will be larger in scale. This has implications for the relative roles of academic and industrial research and the need to coordinate them. The types of large-scale analysis available through gene chips, sequencing of the genome, and proteomics are more amenable to the high-throughput, technician-intensive work of industrial laboratories than to small research laboratories in universities. An example might be the identification of all the kinases, phosphatases, and other enzymes in cells.[9] Industrial research, however, is proprietary and the results not always published, which raises the prospect of duplication of research efforts by government-funded investigators in universities. This increases the need for government and industry to coordinate the parts of big science projects.

We can also expect, in this next decade, to see significant changes in the organization of industrial research itself. The general trend to downsizing or eliminating central research laboratories and decentralizing research units to product divisions was noted above. As a result, industrial research facilities are likely to be more geographically dispersed and culturally diversified, reflecting the growing global reach of investments. The tendency to have laboratories in multiple locations is likely to continue, driven not only by the desire to capitalize on the diversity of skills and creative thinking in different parts of the world, but also by the need to respond to political pressures that tie market entry in a country to a research presence there as well as by market pressures to tailor technologies for individual countries.

Furthermore, as economic opportunities improve in developing countries (some of them are aggressively building up their educational and scientific research capacities), individual researchers will be less willing to emigrate from them. More companies will have research facilities overseas to capture this talent, which will result in even more cultural diversity within companies and a less U.S.-centric view of technology development or, put more positively, greater ability to develop technology systems and products for foreign markets. Communication infrastructure will facilitate more efficient operation of multiple sites and increase opportunities for employees to perform physical experimentation remotely, providing nearly universal access to expensive physical facilities. Research universities are likely to react to this globalization of research facilities and activities by establishing their own global centers of excellence, as the Massachusetts Institute of Technology did when it set up Media Lab facilities in India and Europe.

Companies are likely to become increasingly willing to outsource research or to pool resources with other companies to take on long-term and very large or risky projects to the extent permitted by antitrust regulations. The outsourcing may be to universities or to for-profit research organizations like Battelle, SAIC, or Sarnoff. Pooling of resources can take the form of participation in government- or industry-sponsored research consortia or research centers like Rockwell

---

[9]Enzymes are proteins that speed chemical reactions in the body. Kinases, for example, catalyze the transfer of phosphate from one molecule to another; phosphatases catalyze the hydrolysis of monoesters of phosphoric acid.

Scientific and HRL Laboratories, which are jointly owned by several companies. These strategies can reduce cost, increase efficiency, and shorten time to market.

Companies will also resort, more and more, to creative ways to outsource product development. Acquisition of start-up companies is one way to outsource development and to leverage existing manufacturing and market infrastructure without directly affecting earnings. Venture capital investment in start-up companies, which fosters multiple sources of technology innovation by offering an economic incentive, is an effective way to shorten product development time. In turn, acquisition of start-up companies can improve quarterly profit-and-loss reporting by shifting R&D expenditures into good will. The development of company-owned technologies may also be outsourced in cooperation with venture capitalists, spinning off some technologies for further development by the satellite company or spinning in the developed technology at an appropriate point. The aim will be to match the risk profile of the project and the investment strategies of the company and the venture capitalists. To develop complex and large projects, industry will seek the capabilities of the national laboratories of the Department of Energy. The Intel-led extreme ultraviolet lithography project in cooperation with three national laboratories is an example.

Aside from these broad changes, the field of biotechnology is generating some specific changes of its own, driven by the highly rigorous regulatory environment in which those companies must operate and their deep dependence on strong intellectual property positions. Commercial R&D directed at exploiting genomics for new diagnostics and drug discovery took hold predominantly in smaller or new businesses rather than in large existing pharmaceutical companies. Business models based on near-term development and production of diagnostics were by and large not successful. This created a fertile field for partnerships, mergers, and acquisitions. The entrance of large companies, such as Motorola, Corning, and Agilent, was inevitable, both because of their ability to market and because they possessed cutting-edge technologies relevant to the next generation of chips (e.g., Samsung and chip manufacture and Corning and chip substrates). Currently, almost every kind of relationship exists: partnerships between small and large companies and also between large companies. Superimposed on this dynamic is a large resource drain in the form of a web of legal battles over patents, together with uncertainty surrounding the ultimate resolution of questions about what is patentable in this field.

Finally, industry associations are playing, and will continue to play, an increasingly important role in guiding the direction of R&D by developing industrywide roadmaps modeled after the successful example of the semiconductor industry. The Optoelectronics Industry Development Association is one such association. The Institute of Electrical and Electronic Engineers also continues to play a critical role in the development of industry standards, and even in the codification of de facto standards where technology is moving faster than standards bodies.

PEOPLE

The restructuring of research in the next decade will have ramifications for the scientific and technical workforce and will be affected by a number of changes in that workforce. Education, motivation, and work habits are all likely to be noticeably different in the next decade from the past few decades. Broadly speaking, we will see growing opportunities for scientific or technical entrepreneurship and an increasing need for breadth in the technical workforce—multidisciplinary training within the technical fields and broader training in the management of the technical enterprise.

For example, technical entrepreneurship will continue to flourish as more technical people either form or join start-up companies. And because of the impact of technology on almost every business, there may be more CEOs with technical training. Their success stories would increase the social status of engineers, which would draw more students to technical fields. There is likely to be more interplay between business and engineering schools to encourage entrepreneurship. A student with a technical degree will probably have some familiarity with entrepreneurship, business, and management when he or she graduates and will want to join a company that can provide exposure to business practices rather than take up pure research in academia.

Furthermore, there will be greater pressure to make financial rewards for technical people commensurate with the economic value they create. The reward systems in government laboratories, universities, large corporations, and start-up companies operate under different constraints. To attract and retain star performers, they will need to find ways to provide sufficient compensation, which could make it difficult for organizations that have rigid pay scales to compete for top technical talent. The committee believes that the recent cataclysmic decline of some high-tech sectors is likely to be a temporary situation and that the private sector will continue to be a very strong competitor for talent because of its ability to offer significant financial rewards for scientists in hot fields. Government laboratories are caught in the middle of the bidding war for talent, able to offer neither the financial opportunity of the private sector nor the freedom of the academic environment.

Technical workforce mobility will continue to increase owing to the speed of technology obsolescence and economic cycles. Although individually disruptive, the net result of such cycles is positive because they lead to technology migration and cross-fertilization of know-how. This mobility is being recognized as a major competitive advantage for the United States and is likely to become more widely accepted.

The industrial system will also require new, or at least a greater range of, skills in its technical workforce. For example, shorter time to market is an important competitive advantage. To achieve it, industry will need technical people who are able to manage projects from idea conception to product launch in a concurrent, coordinated fashion. Engineers will have to have managerial skills as well as a clear understanding of all aspects of product engineering, manufacturing, marketing, and intellectual property strategy. Continuing education will become a way of life for engineers to maintain versatility and remain competitive in the job market.

There is an increasing awareness of the shortage of workers with basic technical skills, especially in sectors with rapid business growth. The optics and photonics industry, for example, had and still has an acute shortage of mechanical and optics engineers and optics technicians. Companies have reacted by hiring mechanical and manufacturing engineers from the disk drive industry and trained their own technicians. This speaks to the need for an agile workforce that can be easily deployed in new areas.

The multidisciplinary nature of many of the new technical developments, particularly in the physical and biological sciences, will require that scientists and engineers learn to bridge the communications gap between different disciplines, each with its own vocabulary, jargon, and acronyms. Would-be biological scientists will see substantial shifts in the undergraduate curriculum, with more grounding in mathematics and physics than is now the case. Physical scientists and engineers will need to study biology, a trend we are already seeing, particularly in engineering curricula. For example, at the California Institute of Technology biology is a required course for all students.

Clearly, these new needs will put pressure on universities. While it is unlikely that the traditional departmental structure will change, past practice, particularly in the technical fields, of filling students' programs with required courses in their major will have to give way to new practices, giving students greater latitude and encouraging them to select courses in other departments. This transition may be eased by the increasing need in research for interdepartmental cooperation to be successful in attracting the new kinds of grant funding from either government (the Department of Defense Multidisciplinary University Research Initiative and the NSF National Nanoscale Initiative are examples mentioned earlier) or industry.

Although the fraction of foreign students in U.S. graduate programs in science and engineering has declined during the past few years, non-U.S. citizens on temporary visas continue to make up a high percentage of graduate students and Ph.D.'s. Generally, this appears to be more beneficial than not for the United States. Without this abundant supply of talent, our research universities and technical industry would see dramatic decreases in their global competitiveness. Even those who return to their homeland can enhance our sphere of influence and remain a resource for the United States. Changes, therefore, in public policy on the immigration of foreign science and engineering students to the United States could significantly affect the human resources aspect of the national innovation system.

PATTERNS OF INVESTMENT

Several factors mentioned in preceding sections of this report are reflected in and reinforced by the patterns of investment we can expect to see in the next decade. The desire of large corporations to outsource technological development, coupled with the availability of venture capital and the increasing popularity of technical entrepreneurship, suggests that the proliferation of small start-up companies will continue. This has the significant advantage of creating many parallel points of technology development, which will speed up the technology-to-value-creation process. While there will be many fewer winners than losers, more wealth will be created than lost.

While the proliferation of start-up companies may deplete the technical talent available for basic research, successful technical entrepreneurs can have a major impact on our technology-driven economy. Besides turning technology into businesses, some successful technical entrepreneurs become serial entrepreneurs, starting other companies or investing their profits in other start-ups. Many retain their connection to academia, either returning to universities or making substantial gifts to them. This virtuous cycle is already presaged by the huge private foundations of Gates, Packard, and other early high-tech entrepreneurs similar to those of Carnegie, Rockefeller, and Ford before them.

Investments in long-term R&D by public companies will continue to be negatively affected by the need to achieve short-term financial goals. Companies have always had disincentives to invest in long-term basic research, especially the possibility that competitors could profit from any advances, but the pressure for companies to increase earnings on a quarterly basis has increased the pressure to invest in short-term applied research and development. Central research laboratories that perform basic research and develop infrastructure technology but do not produce short-term, direct financial benefits will probably continue to decline. This leaves gaps in three places: (1) long-term basic research, (2) capital-intensive applied R&D with a long time to return, and (3) critical infrastructure development.

It would be easy to suggest that these gaps must be filled by government funding; indeed, to a great extent that is likely to happen. However, there are complications that will need to be resolved, and the outcome is not clear at this point. First, the eagerness on the part of government, universities, and industry to promote the earliest possible application and commercialization of new science, coupled with what has been a steady natural compression of the elapsed time between fundamental investigations and marketable innovation, has made it difficult to distinguish the kinds of R&D projects that are the appropriate responsibility of government from the kinds that should be supported by the private sector. The growth of government-funded precompetitive consortia, of targeted or highly goal-oriented multiple-institution R&D programs, and of high-risk, high-payoff critical technologies suggests a widening rather than a narrowing of the definition of what is appropriate for government funding. Assuring U.S. success in an era of global technological competition is clearly a factor in this trend.

On the other hand, there have been and will continue to be backlashes. Because these programs are often viewed as helping corporations, there have been very public and often politicized discussions about what constitutes an "American" corporation, making it eligible to participate in these programs, in a world in which multinational corporations are the dominant players. A danger is that excessive restrictions on who can participate in these public programs may actually narrow the group to the point where the approach loses its advantage and the investment is diminished in value.

A second possible backlash, potentially more serious, is that blending the goals of fundamental research with those of technological innovation could offset the tendency of scientists to self-organize for more international cooperation in basic research. If Congress and the public think that a competitive advantage in technology and its application is the same as a competitive advantage in the underlying basic sciences, they may come to believe it is better to compete than to cooperate in basic science. This would not only decrease the cost-effectiveness of global investments in basic research but would also present a major problem for "big science." Projects in astronomy, astrophysics, and high-energy physics, for example, require more rather than less international cooperation to fund equipment and facilities, and pressures to close avenues of international cooperation in R&D might well be a problem in these areas. Indeed, one of the challenges in the next decade will be to develop new and more effective institutional approaches to long-term, multinational funding of research projects.

A third concern continues to affect the public reaction to government funding of programs that blur the distinction between basic research and application—namely, concern about "corporate welfare" or the appropriateness of government intervention in areas where the market should determine winners and losers. The perennial debate over the appropriateness and effectiveness of the Department of Commerce's Advanced Technology Program illustrates the problem well.[10]

A fourth concern is that the increasing emphasis on technological innovation, with its focus on short-term goals, will drive out funding for basic research, which is long-term and not focused on practical goals. The perennial efforts to maintain the basic and applied research parts of the defense R&D budget (the so-called 6.1 and 6.2 accounts) distinct from its development parts illustrate this problem.

The second gap area mentioned above—capital-intensive applied R&D—

---

[10]For an overview of the issues, see the recent report of a symposium on the Advanced Technology Program: National Research Council, Board on Science, Technology, and Economic Policy, *The Advanced Technology Program: Challenges and Opportunities.* Washington, D.C.: National Academy Press, 1999.

appears likely to start at least one trend that runs counter to the proliferation of small start-up companies competing to produce new technologies. Consolidations in the aircraft industry and defense contracting exemplify a countertrend driven by capital constraints. In the information technology area, we may see a similar phenomenon. The huge difference between the investment needed to be a player in the silicon arena versus that needed in the optical communications arena may one day become an important factor. Costs for a new silicon plant are now so large that it will be increasingly difficult for more than one player to be active, whereas the optical arena may stay highly competitive.

The third gap mentioned above is the need for critical infrastructure. Here the role of government is clear and likely to grow in importance. Advances in metrology and the development of technical standards are critical infrastructure capabilities that must be in place to facilitate the efficient incorporation of innovations into the economy, whether the innovations are in the biological or physical sciences. The need for such infrastructure is even greater when the innovations come from small companies, which do not have the resources large corporations have to fund large research facilities and sophisticated equipment.

The focus on short-term financial goals might be offset in part by new corporate approaches to analyzing R&D investments, such as real options methods. The justification for investment in R&D, including basic research, might be increased by new methods of allocating internal investment within firms that take more account of the role of R&D investment in corporate innovation.

PUBLIC POLICY ISSUES

There is every indication that the United States will continue to recognize entrepreneurship as a competitive advantage for the country and will continue to promote it. It is likely, moreover, that the federal government will continue to encourage investment through tax policy—for example, tax credits for R&D, R&D partnerships, and preferential capital gains tax.

Although the United States has generally been sensitive to business interests in the area of regulation, changes in industrial organization, the global character of the economy, and rapidly emerging technologies will present a number of new challenges for the regulatory system. We are already facing some major questions related to antitrust regulation, illustrated by the government's prosecution of Microsoft for violation of antitrust statutes and the United States-European Union controversy over the proposed GE-Honeywell merger.

The Microsoft case arose because the usefulness (and, consequently, the economic value) of networked systems strongly depends on the number of connected nodes in the network (the usefulness is approximately proportional to the square of the number of nodes). When competition reduces the number of nodes, or users, of a particular system, it also decreases the value of the system. As a result, some economists now argue that the public interest would be best protected by a sequence of temporary monopolies rather than by a forced sharing of the market at every moment. Therefore, there is likely to be pressure in the coming years to design antitrust regulations that prevent the long-term perpetuation of information network monopolies rather than regulations that prevent their formation altogether.

The GE-Honeywell attempted merger raised another issue—the difference in views in the United States and the European Union over whether mergers that allow horizontal integration across business or technology sectors violate antitrust principles by providing the merged entity with the ability to gain excessive market share within one or more of the individual sectors of which it is comprised (which was the position of the European Union). In view of the global trend to mergers and acquisitions, discussed in an earlier section, the ultimate resolution of this issue will have a significant effect on the strategy of multinationals in the next decade.

Still another important regulatory issue has been introduced by the growth of the medical technology industry, including biotechnology, tissue engineering, and medical devices. Although the Food and Drug Administration (FDA) has had responsibility for assuring the safety and effectiveness of medical devices since the mid-1970s, most observers would argue that the regulatory procedures and institutional organization have not evolved adequately from the days in which the FDA's responsibility was almost entirely restricted to drug assessments. Since the development of medical devices follows an entirely different path from that of drugs, the medical devices industry needs very different assessment procedures, more room for design iteration, closer attention to systems issues, and different measures of performance.

The next decade is likely to further complicate the matter of regulation as the new technologies mature. For example, a hybrid pancreas that consists of a polymer matrix protecting and supporting cultured human beta cells that produce insulin for the control of sugar level is presently viewed by the FDA as a device, a biologic, and a drug, each of which is subject to regulation by a separate division of the agency under different protocols. If the United States is to be able to take advantage of the rapid, multidisciplinary innovation that is bound to occur during the next decade, there will be great pressure to develop a regulatory system that can evolve as quickly as the technologies it is charged with evaluating. One likely development is the evolution of the present system, in which medical devices are categorized as either "experimental" or "clinical," into a system that recognizes a continuum of development from prototype to mature technology.

Medical devices are one area in which the nation's regulatory procedures play an important role in determining its competitiveness. Medical technology companies have been moving a substantial fraction of their R&D to Europe because they view European device regulation as more rational and efficient and more conducive to technological innovation. On the other hand, more accommodating systems for regulating genetically modified organisms has kept the United States in the lead in this area.

Intellectual property law, in both copyrights and patents, is another area in which technological advances are straining the system. Here, the choices made during the next decade will have a strong influence on U.S. competitiveness. For example, the new opportunities that computers and communication technologies offer for adding value to public data and information by reorganizing them and mining them are giving rise to an explosion of copyrighted databases. The government will face pressures to cut back on its efforts to refine raw data, exploiting the same technologies to make the data easier to use, and it may even be less aggressive or less accommodating in making raw data easily available, in order to avoid competing with the private sector. In the long run, this could work against the interests of small companies, which are less able to absorb the costs of accessing the information they need.

A related issue, whose resolution during the next decade will have an influence on scientific research, concerns the doctrine of fair use—the allowable, free copying of copyrighted information for a set of specifically designated purposes. In the past, fair use represented a compromise in which the owners of copyrighted information, recognizing the practical limitations on their ability to monitor limited copying and the relatively insignificant consequences of such copying, accommodated the needs of scholars and some others. With the advent of new information and communication technologies, monitoring has become practical and copying more problematical because of the ease with which enormous numbers of copies can be made and distributed. As a result, there has been great international pressure (through the World Intellectual Property Organization) to severely reduce fair use. The concern, expressed by a number of scientific and professional organizations in the United States particularly, is that this would limit the exchange and use of scientific data. More work is needed in the public policy arena to find a resolution to this issue that preserves the openness of communication in science, a vital aspect of scientific progress.

The tensions between the openness of science and the need to protect intellectual property in order to encourage innovation also reach into the area of patents, especially in biotechnology. The patenting of genes and even DNA sequences is clearly an unresolved issue, with the United States and the European Union diverging in their positions. The U.S. Patent Office has been fairly open to the patenting of genes, even where specific gene function has not yet been established. In fact, it has allowed patentees to make claims to future discoveries of gene function. Some have argued that this creates a useful incentive to companies to move aggressively to sequence genes, but to many the more likely effect is to create a disincentive to investigators to pursue other applications of a particular gene as the several proteins it expresses are discovered and understood.

Current U.S. patent and copyright laws may inhibit the development of new businesses, because disputes are usually resolved through the courts rather than by the Patent Office. The cost of litigation is very high and therefore favors established companies. In some new technological areas, such as biotechnology, the lack of case law to help interpretation is especially problematical, because the uncertainty may inhibit investment in innovation. In others, patenting is less important than alternative forms of protection, such as trade secrecy.

As technological innovation shifts to small or medium-size companies, brand recognition becomes a less effective way of building consumer confidence and facilitating the adoption of new products and services. That will increase the role of the government as an honest broker, providing and disseminating independent assessments. The Internet is an important tool for disseminating assessments, but the vast array of information of varying quality that is available on the Internet makes it all the more difficult to establish credibility. The government may need to certify data on the Internet, especially where the information is widely used as the basis for research, product development, or manufacturing.

Even if the validation of data is judged to be an inappropriate role for the government, there is a need for formal data format standards that can address both longevity and migration issues. In addition, there is a need for standards in the presentation of genomics/proteomics sequences (in addition to the need for standards of sequence accuracy, noted earlier in this report) that is not currently being met by NIH, FDA, or NIST.

Finally, it should be noted that in the next decade, global standards will take on increasing importance for products in all areas. In much the same way as data and communication standards in telephony, fax machines, and the TCP/IP protocols of the Internet facilitate the development of global networks and the easy sharing of information, standards will be needed for many other products. Product standards can, it is true, serve as trade barriers, but for a nation committed to free trade on a global scale, they can also facilitate trade and help to gain competitive advantages. The failure of the United States to gain a significant share of the global market in cellular telephony is a good example of the price of not moving quickly. Today, product standards are voluntary in the United States but are sometimes import requirements in other countries. It may well be that as this field becomes more important, federal government will need to play a more direct role, especially as it is the critical voice in the World Trade Organization and other international venues where these issues are negotiated.

4

Pull Factors

Historians of science and technology have argued over the extent to which push factors—the drivers of new scientific knowledge and human inventiveness—determine the directions of technological development. Some, the hard-line technological determinists, believe that in one way or another, push factors alone ultimately determine which technologies will be adopted and, as well, what social changes will occur. Others, however, are less convinced that the matter is so simple. Their position, sometimes called soft determinism, is that technological development and adoption involve a more interactive process. Demand in the economic sense, based on consumer preferences, is considered to be very important in determining whether a technology will succeed. The AT&T Picturephone is a classic instance of a technological capability that consumers were not interested in having, in part because of its high price but in part because of its impact on privacy. The societal context of an age also has an influence on the scientific questions that are asked, and values, biases, needs, markets, legal structures, and other circumstances determine what research will be supported and which projects will turn out to be acceptable or—even more important—viable.

The committee is in the latter camp and believes that, in the next decade, societal (or pull) factors will clearly have an important influence. Among the factors that will place new demands on U.S. science and technology and offer new opportunities for U.S. industry are the following:

- The need for greater efficiency and effectiveness of supply chains and of services,
  - Aging of the population,
  - Rapidly accelerating urbanization worldwide,

- The revolution in military affairs,
- The quest for alternative energy sources,
- Widespread global poverty,
- The quest for global sustainability, and
- Protection from natural and man-caused disasters (terrorism and war).

Our ever-increasing ability to alter biological forms and to change what many would view as the natural course of life is challenging personal beliefs, raising deep ethical questions, and leading to society-wide discussions about whether we *should* do all the things we *can* do. Our ability to gather, store, manipulate, and communicate digitized information is shifting power between and among government institutions, nongovernment organizations, the private sector, and individuals in ways that can alter the political and legal structure of the society. Many of these capabilities impinge on individual rights or violate strongly held beliefs, and as the public becomes more and more aware of the potential consequences, it is pressing for a voice in determining just which technology-driven changes are valuable and should be allowed to go forward and which should be proscribed or controlled by appropriate legislation.

The market, of course, continues to be an important factor in what innovations succeed, but it is driven by what people value. There is evidence, for example (discussed in more detail below), that what people value in health care is changing and that this will have an effect on which medical technologies are acceptable. The explosion of interest in and concern about the environment, a fringe issue just 30 years ago, appears likely to continue and to affect what people look for and insist upon in technological products and processes. Indeed, beyond market factors, there is every indication that there will be more, rather than less, environmental regulation in the future and this, too, will affect the direction of technological development.

INTERNATIONAL CHALLENGES

At the international level, the United States faces several challenges that will demand contributions from science and technology. They include the asymmetric threat of terrorism and the economic globalization of economic activities, including R&D.

Antiterrorism

The United States must provide for its security in the face of challenges quite different from those of the Cold War, a situation reinforced by the terrorism acts of September 11, 2001, which expanded national security needs to include measures for homeland security to an unprecedented degree. Short-term problems include improved security for the mail system, airlines, nuclear power plants and other key infrastructure facilities, and prominent buildings. The new security requirements cry out for better technologies for measurement and analysis, needed to improve detection and identification of biological, chemical, radiological, and nuclear threats, and for better ways to destroy or neutralize dangerous biological and other substances used as weapons. For example, "labs on a chip" and biochips could be developed that identify known chemical and biological warfare agents within seconds. Miniature robots with sensors and scanners could sweep sites to detect chemical, biological, and nuclear hazards or to locate terrorists or victims in collapsed buildings. Also needed are better methods of identification, such as national identification cards and biometric identification technologies.

Over the long run, there will be a need to improve the protection of key national infrastructures, including computer networks and other communications systems, power grids, and transportation systems, and the ability of buildings and other facilities to withstand structural damage from attacks. Antiterrorism products purchased by state and local governments as well as the private sector will need to be tested and validated for effectiveness and reliability, and standards for interoperability will need to be developed. All this will have to be achieved in the face of changes in the conduct of research itself, including improved security measures for classified work, increased scrutiny of science and engineering students and researchers, especially foreign students and researchers, and limitations on the sharing of some scientific information with scientists in other countries, all of which impinge on the tradition of open dialogue among scientists, which has always hastened progress.

Globalization

The United States must also preserve its economic strength in the face of increasing global competition, particularly from developed countries. At the same time, in the interest of preserving global stability, on which its own security and well-being rests, the nation must do what it can to promote and assist economic growth in the developing world. Technology, of course, figures prominently in the strategies for achieving U.S. goals in all these areas, so that it is to be expected that the government will use its market and regulatory powers to encourage technological development along trajectories that serve these goals.

In addition to the globalization of trade, there is the globalization of R&D. Companies around the world, including U.S. companies, have been locating more R&D abroad, to help them meet the needs of foreign customers, keep abreast of advances in science and technology, employ the talents of foreign scientists and engineers, and work with foreign R&D laboratories. This trend raises a number of issues, not least the appropriate role of the U.S. government not only in supporting industrial research at companies that are becoming more international than national in ownership and location, but also in setting industrial standards and regulations for them.

In the following sections, the potential influence of these pull factors is discussed.

BIOLOGICAL SCIENCE AND ENGINEERING

It is not unusual for science and technology to raise value issues for people. But the modern biological sciences are unique in that, for many people, our ever-increasing capacity to understand the structure and function of living systems at the most fundamental level and to manipulate and modify both structure and function raise questions about the meaning of being human. It is one thing to cure a disease; it is quite another to replace an organ actively involved in maintaining life. And it is qualitatively different still to modify the building blocks and processes of a living system.

The issue is particularly complicated because it goes beyond the relatively narrow questions of just how much the biological system is being altered or what broad effect the alterations might have on the nature of the organism—more or less factual matters. For many, the important thing is the symbolic meaning of these new technologies, such as cloning, and what our willingness to use them says about our values.

For others, the issue relates either directly or indirectly to religious beliefs about the beginning and the end of life—and also raises the question of what those terms mean (a question that has been debated for a few decades now). And even for research whose goals are not controversial, as they were for research using fetal tissue, its possible relation to an issue of great concern—abortion—is sufficient to enmesh the research with that other issue.

Still another complicating factor, particularly in Europe but also in the United States, is the uncomfortable history of the eugenics movement. It has led a number of people to be concerned that genetic engineering will reinforce unhealthy normative notions of human perfection, leading to the view that people with disabilities or whose physical appearance does not conform to those norms are inferior. It is, therefore, no surprise that organizations representing people with disabilities have been particularly vocal in opposing most aspects of genetic engineering.

It is the view of the committee that the great promise of molecular and cellular biology will overcome these objections, but there are some significant caveats. The development of clinical therapies based on molecular and cellular manipulation will be subject to extremely close scrutiny and held to an evidentiary standard and a demonstration of efficacy even more rigorous than usual in medicine. Indeed, there is much work to be done to convince the public that genetic alterations have only the intended effects and cannot lead to inadvertent alterations. Our relatively complete knowledge of the human genome (and the genomes of other species) should help in that respect.

It is certainly likely that standing advisory committees with significant representation from the public and the bioethics community will exercise major influence in determining which lines of research are allowed and which are proscribed. For example, it is very unlikely that any therapies based on germ cell modification will be allowed, even if scientific research leads to greater confidence that such alterations can be carefully prescribed and controlled. Of course, the ability of the government to impose research standards and to regulate research is much greater when it supports the research, as was made clear recently in the case of embryonic stem cells. It seems likely that this will provide a strong incentive for the government to stay involved in such research.

Genetically modified organisms (GMOs) in agriculture are likely to encounter even stronger resistance, both domestically and internationally, than genetic manipulation in humans, primarily because the benefits to society are seen as less important than advances in curing human diseases. Thus, stronger weight will be given to arguments against the genetic modification of plants. The greatest challenges are to the use of genetic manipulation to build "nonnatural" genes into plants—for example, to program plants to produce herbicides and pesticides. Politically, these changes give rise to a coalition between environmental activists and those who oppose genetically modified foods. Our ability to develop and agree on approaches to field testing and ecological assessment will clearly be very important, as recent examples of the spread of engineered genes from Bt corn (a type of GMO) to nonbiotech corn plants in the field indicate. The related issue of mixing of a GMO (Starlink corn) into snack foods is another area with implications for public support of technology development in this area. Continued growth of agricultural biotechnology will depend on satisfactory solutions to these challenges.

The use of genetic engineering techniques as a more efficient approach to traditional plant breeding for optimal characteristics will be somewhat easier to sell. Of course, a significant fraction of U.S. crops has already been developed using these techniques. They should become even more widely acceptable, at least in this country, as our knowledge of each plant's genome allows us to ensure that no changes but the intended changes have taken place.

At the international level, the problem is more difficult because it is hard to separate legitimate scientific differences between countries from trade issues. If genetic alteration does not take hold in other parts of the developed world, as it has not to any great extent so far, there will certainly be great pressure in other countries to exclude American GMOs and to cast the issue as one of public health standards. This reinforces the need for objective procedures for establishing scientific fact at the international level when important public policy rests on the outcome.

There are similar differences with respect to medical technologies—namely, the United States and the European Union have markedly different approaches to regulating medical devices. Generally speaking, and in strong contrast to its views on genetic manipulation, the European Union has been much more receptive to new medical technologies and, in the view of most medical device companies, more successful than the United States in developing protocols for device assessment that are both effective and efficient. The early indications are that medical device companies are willing to move research and development operations to Europe and to introduce new technologies there at a much earlier time. This suggests that there will be pressure from the U.S. public in the next decade for the United States to move closer to the European Union in this matter so that Americans do not lose their lead in medical devices or the health benefits that derive from them.

Two other issues related to medical technologies will grow in importance in the next decade. First, driven in part by their general discomfort with what might be termed the technological invasion of their bodies, the public will look for technologies that increase their personal control over their medical care rather than those that make them more dependent on physicians and technicians. Computer-aided devices that they are able to regulate without the intervention of a physician, diagnostic kits that they can use on their own, and interactive medical sites on the Web that can help them to make informed decisions about their own treatment: all are more likely to find wide acceptance. There will be a premium on technologies whose sophistication is applied to create simplicity for the patient.

The other pressing issue is the growing cost of health care in the United States—it now exceeds $1 trillion per year, more than 14 percent of GDP—and the fact that more than 40 million Americans lack health insurance. The introduction of medical technology has often cited as one cause of this increase in medical costs. How much of the increase is due to technology and whether that increase is cost-effective (that is, whether it cures particular diseases or extends human longevity at a lower cost than existing treatments—if such treatments exist) are matters of continuing debate. But the concern about total cost and the fact there is likely to be differential access to new technologies because health insurance is not universal present difficult social problems that will undoubtedly affect the rate of adoption of medical technologies in the next decade. For example, if chip diagnostic manufacturers want to compete in the $19 billion diagnostic market, the assays they develop will have to be inexpensive.

COMPUTER AND INFORMATION SCIENCE AND TECHNOLOGY

Biomedical trends help to illustrate a key issue in the computer and information field—privacy. Health care and information technology intersect in a number of ways, but the area of patient records is a particularly noteworthy synergy because our ability to get more data about an individual is growing as rapidly as our ability to store, analyze, and transmit those data. Obviously, the richer the information the health care system can gather about a person's medical condition, and the more it can analyze and synthesize that information, the more effectively it can preserve the health of the individual.

However, others could use this same information to the disadvantage of the individual. For example, if a prospective employer or an insurance company were aware of a person's genetic predisposition to heart disease or some types of cancer, it might refuse to employ the person or to provide coverage.

This hypothetical example raises two major issues. Because this kind of adverse selection would be applied to a person who has no more than the *potential* for a certain disease,[1] it would significantly extend the kind of adverse selection traditionally applied by insurance companies to people with preexisting conditions. Many would argue that such deselection would compromise an individual's entitlement to equal protection under the law. Second, if information about a person were obtained from easily accessible electronic records without the person's permission, it would be a significant violation of his or her privacy.

Both issues are important, but it is the latter—the privacy issue—that is likely to present the toughest policy difficulties during the next decade and to have the most impact on the development and use of information technologies in uses well beyond health care. Privacy must be understood not so much as the confidentiality of information about an individual as the right of the individual to *control* information about himself or herself, including being able to decide to keep it confidential or to share it selectively.

At the moment, U.S. law concerning privacy is in a state of flux. There is no one legislative authority, jurisdiction, or law governing privacy. There are both state and federal laws, but only for certain aspects of privacy. There is a law protecting the privacy of an individual's record of videotape rental, but none covering his or her Web site visits. There is discussion about not allowing a person's genetic data to be used in reaching decisions about insurance coverage but no similar discussion about using family history for the same purpose. There are regulations concerning the release of data held by the federal government about individuals, but much more limited regulation (at least in the United States) covering what private entities are allowed to do with such data.

It is clear to many people that sharing data about oneself may often be in one's interest—and the technologies being developed will vastly increase that advantage and the services available by the sharing of data. Emergency medical information, global positioning systems, personal preferences in books and mu-

---

[1] There is an important distinction to be made here between genomic analysis and proteomic analysis, which will become important as the capacity for proteomic analysis increases in the next decade. The genome only indicates a predisposition to a disease. Whether it actually occurs will also depend on environmental factors to which an individual is exposed and, indeed, on luck. Protein expression, however, is usually associated with the occurrence of a disease or its incipiency and presents a quite different situation.

sic, information mining services, entertainment, interest group linkages, all hold out the promise of leveraging technology for personal benefit. However, unless we, as a society, can develop an orderly and rational approach to the protection of privacy (and the limitations of that protection), the technology development will be hampered by legal uncertainties, and individuals will see its threats more clearly than its promises.

In reaching some general understanding of the social issue involved, it is useful to frame the problem as a discussion of the appropriate boundary between public and private spaces. Clearly, there should be some rational boundary. People are entitled to control their own lives (and information about themselves), but in return for membership in the society and all that such membership entails, they take on obligations that may place limits on their privacy. That is part of what may be thought of as an implied social contract, given immediacy by the events of September 11.

Unfortunately, our decisions on where that boundary properly belongs have either been based on the political or security exigencies of the moment or been driven by the current state of the relevant technology. For example, as heavier encryption technologies become available to individuals, private space widens because individuals are better able to shield information or transactions from government or other entities. As government signal-capture and computational capacity increase, private space narrows.

Privacy is only one of the many social, political, and legal issues that will affect how Web-based information technologies develop in the next decade. Free speech will conflict with the desire to protect minors and with concerns about hate speech. The historically libertarian nature of the Internet will conflict with the desire of governments to hold Internet hosts or other providers accountable for enforcing local ordinances. The Internet's openness and its historical and deliberate lack of governance structure will also conflict with the desire of governments to provide consumer protection for those doing business on the Internet, to enforce intellectual property laws, to deal with taxation issues, or to resolve jurisdictional issues.

Because the Internet is global, it is to be expected that issues of harmonization will be very important in both governmental and commercial terms. For example, as the Internet penetrates other societies, particularly those in less-developed nations, the issue of software language will become increasingly important, and it is likely that there will be pressure on almost all software developers to have source code that is compatible with a variety of front-end language systems. TCP/IP is, of course, established throughout the world as the standardized packet-switching system, and it is likely to continue as such through the decade. However, although the routing of messages is, in principle, random, the network hardware architecture is such that most messages actually pass through the United States. That situation is likely to be increasingly uncomfortable for many nations and may bring pressure to diminish the de facto hub dominance of the United States.

ENVIRONMENTAL ISSUES

Interest in the environment has grown enormously since the United Nations Conference on the Human Environment, held at Stockholm in 1972. What was once a peripheral issue of concern to only a few groups has become a mainstream issue in all countries of the developed world, including the United States, and the list of concerns included in its compass has vastly expanded. Questions remain as to how deeply committed the public is to taking action to protect the environment, but several factors suggest that the next decade will see increasing concern and deepening commitment to give environmental protection a high priority.

First, the growing interest has given rise to new scientific investigations (and even relatively new sciences, such as ecology), which have reinforced public concerns. Observational data on the effects of chlorofluorocarbons on stratospheric ozone led in almost record time to an international treaty to eliminate their use and simultaneously stimulated the development of effective substitutes for them. Indeed, work continues on still better materials than the hydrofluorocarbons now being used. The work of the Intergovernmental Panel on Climate Change and the National Academy of Sciences points increasingly to anthropic causes of global warming and the seriousness of the consequences.

Second, improved instrumentation has led to orders of magnitude increases in our ability to detect trace compounds in the environment. This, in turn, has given rise to pressures to remove those compounds, to prevent them from entering the environment in the first place, or to improve our understanding of dose/response effects for various potential toxins in order to establish a rational basis for setting tolerance limits.

Third, the correlations among gross world product, energy use, and environmental stress create significant tensions between the justifiable desire to improve the income and standard of living of people all over the world and the equally justifiable desire to protect the environment. Some argue that technology is part of the problem, but the committee believes that it is more likely to be part of the solution—and the pressure to make it so will help to guide technology development over the next decade. Technologies will be encouraged that accomplish the following: decrease energy intensity (the energy used per unit of gross product); move us away from fossil fuel-based primary energy production and/or increase the efficiency of primary energy production, transmission, and distribution; reduce the concentrations of greenhouse gases in the atmosphere; cut down on the material flows associated with production; prevent the dispersion of toxic wastes from point sources; reduce the need for water, fertilizers, herbicides, and pesticides in agriculture; or provide for the economical remediation of polluted land and bodies of water.

The increasing pressure to incorporate technologies that protect the environment into all production activities will give large companies a competitive advantage because they are much more able to conduct R&D across a broad spectrum of fields. Small companies, which—as noted earlier—are often the source of innovations, have fewer financial resources and must choose between spending their R&D dollars on current product or on innovation. Thus, absent significant government investment in environmental technologies, the pressure for environmental protection will generally work against small and medium-size companies unless their business is environmental technologies.

On the other hand, raising the standards for environmental protection in the United States—given the ability of the country's R&D establishment to respond to such a stimulus with increased innovation in environmental technologies—may well give U.S. companies a comparative advantage internationally and lead them to push the government to negotiate the harmonization and tightening of environmental standards throughout the world. Thus, the corporate sector itself may become a force for increased environmental protection.

All of the fields identified in this report have already had significant impact on environmental issues and are likely to figure heavily in future efforts to protect the environment. Computer technology has been a key to the global modeling studies that are confirming society's contribution to the greenhouse effect. Remote satellite observations have provided an enormous amount of retrospective and prospective data on the consequences of warming and the spread and effect of various pollutants. Advances in sensors as well as in remote imaging have played an important role in monitoring compliance with international treaties such as the treaty governing transboundary air pollution. Microbial approaches to cleaning up oil spills were shown to be effective in a number of highly publicized incidents. New materials have decreased our dependence on natural resources; for example, fiber optics has limited the rate of increase in demand for copper, the mining of which has severe environmental consequences. Strong lightweight materials used in transportation vehicles have helped to limit the increase in energy consumption in that sector.

In the production of primary energy, computer design and control of windmill blades has so lowered the cost of wind-generated energy that it is competitive with fossil fuels in many parts of the country. Stationary fuel cells are also in use and are perhaps the most promising new technology both for primary energy generation and for transportation. Both technologies are likely to grow in importance in the next decade.

This record of success, coupled with the likely increase in emphasis on environmental protection in the next decade, suggests that there will be strong pressures—and rewards—for companies to exploit these fields where scientific knowledge is rapidly changing to develop innovative technologies for protecting the environment.

INTERNATIONAL SECURITY AND THE GLOBAL ECONOMY

In recent decades, the once predominant concept of a linear progression from basic research to innovation has given way to the realization that there is often significant interaction between basic research and technological development. It has become more and more difficult to distinguish basic from applied research; the time frame from new idea to application has been foreshortened; feedback loops have been created, with technological development providing data and insights that inform new rounds of basic research; and cross-fertilization between different fields of science has made it all but impossible to judge what field of research will provide the breakthrough for a desired innovation. In short, the linear model of R&D has given way to a matrix-feedback model.

The shift is important for several reasons. First, the strategic management of R&D has become significantly more difficult, because it is much harder to determine what research needs to be done in order to achieve some technological goal. Second, the clear demarcation between activities whose support is the appropriate responsibility of government and those that should be funded by the corporate sector has blurred, as has the line between precompetitive and competitive research. Third, in a world marked by the globalization of both research and technological development and the linking of economies, nations need to cooperate in some aspects of R&D and compete in others. The patterns of R&D development make it increasingly difficult to say which activities should be cooperative and which competitive.

The international cooperation/competition dichotomy could have important consequences for research because it can lead to either underinvestment or overinvestment in particular research areas. If cooperation amounts to little more than the open availability of new scientific knowledge to the entire global community, nations may assume that they will be able to benefit from advances anywhere in the world and need not make any investment themselves. Economists point out that this free rider phenomenon can lead to global underinvestment in basic research. On the other hand, if competition is the dominant theme even at an early stage in the R&D cycle, overinvestment could result: Nations would compete for early market advantage by embarking separately on research, thereby duplicating one another's efforts. Since only one nation will ultimately gain the advantage in a particular area, the investments by other nations will have been wasted.

The growth of scientific and technological capacity throughout the developed world will certainly continue in the next decade, and slow but steady growth can also be expected in certain of the newly industrialized nations. Therefore, we will be continually need to find ways to share the costs of R&D and to share technical talent at the same time as we work to achieve our national economic goals.

It seems more likely that, if we err, it will be on the side of failing to exploit the potential for cooperation. Although the United States continues to be the foremost contributor to the world's research literature, with about 35 percent of published research papers having American authorship, other countries now contribute more than half. The long experience of this country as the dominant player in R&D has resulted in an underemphasis on developing the language skills and the perspective necessary to learn from others. Added to that problem, political concerns about losing technological advantage by allowing others to ride free on our R&D advances will probably continue to lead to sporadic efforts to protect our scientific and technological knowledge base. This has been manifest in occasional efforts to limit foreign corporation participation in consortia or to proscribe foreign student participation in government-sponsored research projects. It remains to be seen whether increased efforts to facilitate technology transfer between universities and industry will exacerbate that problem, but it is not unreasonable to expect that it might.

The emergence of astronomy, astrophysics, and particle physics—"big science"—has added impetus to the need for international cooperation to share the cost burden, but big science, too, will continue to encounter obstacles, some of which originate with scientists themselves. Since a facility must be sited somewhere, scientists as well as governments are usually eager to have it in their own country for reasons of pride and convenience, and siting it elsewhere is considered a negotiating loss. It will be interesting to see whether advances in information technology in the next decade reduce the value and cachet of a country's possessing large experimental facilities, encouraging more cooperative big science projects. It will also be interesting to see whether the inability of our superconducting supercollider to attract international sponsorship will lead us to find better ways of seeking international cooperation early on in the planning of large facilities in order to create an atmosphere of partnership.

A complication related to this last point is that mechanisms for international financing of research are not well developed, particularly where multiyear efforts are involved. The United States, for example, has no easy way of guaranteeing multiyear support for projects because its budget cycle is annual. Two frequently mentioned solutions are (1) to use treaties to formalize research arrangements since treaties are one of the few ways for the U.S. government to obligate future funds and (2) to have the U.S. government delegate authority and responsibility for multinational research to quasi-governmental institutions.

Finally, there is the question of the relationship between technological development and national security, an issue given great urgency and prominence by the events of September 11. It is a problem made considerably more difficult by the emergence of dual-use technologies and the shift from "spin off" to "spin on" in the relation between the technologies for military and civilian purposes.[2]

---

[2]"Spin off" refers to the commercial use of technologies originally developed through mission-driven research funded by the Department of Defense, the Department of Energy, and the National

Much has been written about the shift from spin off to spin on in technology development. For much of the Cold War, the most sophisticated modern technologies were developed for military applications with support from the government. One of the justifications for the investments was that the technological innovations were ultimately spun off to civilian applications that improved the lives of the general citizenry. The Internet, the Global Positioning System, satellite imagery, composite materials, even Gatorade, came from what were once military applications.

But the very success of these examples of technological diffusion created huge markets and industries that stimulated civilian R&D and led to the current situation, in which the most sophisticated technologies are being developed in the civilian sector and later subsequently adapted for military use. This has required a sea change in the culture of the military technology establishment, which is well under way. It has required rethinking the issue of technical specifications to better utilize off-the-shelf technologies. It has somewhat altered the traditional prime contractor system of the Department of Defense. It has led to a heavier reliance on civilian infrastructure in communications. And it has increased reliance on non-American suppliers (along with altered strategies to ensure the stability of supplies under a variety of circumstances).

There is much that is positive about these changes, but the emergence of technologies that can serve both military and civilian purposes—dual-use technologies—has also created difficulties. The global diffusion of technologies for civilian purposes also makes those technologies available for military purposes. The early strategy for dealing with the problem was the negotiation (with our trading partners) of a list of technologies that were not to be exported to designated countries—the so-called COCOM list.[3] Heavily represented on the list were computer and other information technologies, but also certain advanced materials and other emerging technologies.

To many in the corporate sector, the restrictions created a disadvantage for U.S. industry, so with the end of the Cold War, the list of prohibited items was renegotiated and considerably reduced in scope. Greater emphasis was placed on restricting the distribution of systems used mainly, if not exclusively, by the military, accompanied by a loosening of restrictions on the underlying technologies. Insofar as this has made it considerably easier to develop global markets in information-related technologies and in advanced materials methods, the U.S. economy has benefited.

Aeronautics and Space Administration. "Spin on" refers to the adoption of commercial technologies by the military and other mission agencies (also called dual-use technologies).

[3]COCOM stands for Coordinating Committee for Multilateral Export Controls, an organization sponsored by the United States and its allies.

In the current situation, however, the threats the United States and its allies face are less from sophisticated technologies adopted and used against us by state entities than they are from clandestine groups using technologies of modest sophistication. Under these circumstances, distinguishing equipment and technology of military use from those with a legitimate civilian purpose is much more difficult, particularly where biological systems are involved. The fermenters for growing useful microbes and pharmaceuticals look much the same as those for growing dangerous species.

Moreover, the threat posed by the possible misuse of genetic engineering to create highly virulent species is likely to increase public distrust of the use of those same techniques to improve the quality of food, the hardiness of plants, and the yield of crops. Therefore, if the threat of biological terrorism grows in the next decade, the application of genetically modified organisms for other purposes is likely to be severely restricted.

In sum, the challenge posed by dual-use technologies in the next decade is likely to be considerably different from the challenge in the past. The threat remains that sophisticated weaponry technology—for example, long-range, remote-guided missiles carrying nuclear warheads—will be misused; recent acts of terrorism remind us of the need to develop better defenses against low-tech, poor-men's weapons, including conventional explosives as well as biological and chemical weapons. For the scientific community, the challenge will be to develop approaches for recognizing, preventing, and combating the misuse of these dual-use technologies. Indeed, the ability to do so may turn out to be a necessary element in their widespread adoption.

There have already been some rudimentary efforts in this direction. For example, some experts have studied means of ascertaining the origin of chemicals or biologicals to help in tracing material while others are working on new sensors for detecting recent shifts in the use of fermenters, and political/economic programs have been initiated to minimize the likelihood that underfunded or underemployed scientists in the former Soviet Union will be tempted to cooperate with terrorist groups. More needs to be done, and it is likely that rather than separating military research from civilian research, a dual perspective will become a normal and continuing requirement for those working on a wide range of technological applications.

5

Conclusions

The committee was not asked to recommend policies to NIST but to report its observations and, where reasonable, findings and conclusions on trends in the institutional, social, economic, and political environment in which the science and technology enterprise exists. The text of its report consists of a series of observations, findings, and conclusions about particular fields of research and areas of technology. Still, analyzing possible futures for science and technology by looking at push, contextual, and pull factors, while analytically powerful for the purposes of this report, tends to underplay the overarching themes, which are discussed in this final chapter.

- Although it is not possible to forecast what specific advances will be made or when, progress in science and technology will continue to be extraordinarily robust, offering substantial benefits to society and expanding opportunities for further progress.

The report includes many examples of promising research advances and technological developments. Many of the most significant advances stem from the intensifying and synergistic convergence among nominally disparate fields. As one example, materials science and information technology have played a key role in the development of DNA microarray "chips," which have made rapid genetic screening possible. As another example, we seem to be at the beginning of an increasingly sophisticated understanding of self-assembly of large biomolecules that might lead to new applications in computational systems.

- The amount and direction of research and technology development are shaped by the environment or the context in important ways, including governmental investment, tax, and regulatory policies; institutional arrangements; and social values. Public and private decisions on funding R&D, including projects, facilities and equipment, and education and training, influence what will be pursued and how fast. Indeed, there are indications that public views on the appropriateness of certain research directions will be reflected not merely in funding decisions, but also in direct regulation. Current law, for example, proscribes the development of software to circumvent certain commercial encryption systems, and it seems possible that certain kinds of human cloning may soon be proscribed.

- The institutions that encourage, support, and regulate both research and technological innovation will be increasingly challenged by changes in industrial structure, management, and financing of R&D; by the increasingly global nature of technological innovation as well as the economy; and by social pressures that will affect the legislative agenda. On the one hand, the challenge will be to design antitrust, tax, intellectual property, and capital formation policies to reflect how technical innovation actually happens. On the other hand, the agencies and institutions, both public and private, that fund and provide various kinds of infrastructure to support innovation—from educational institutions to standards developers to funding agencies to those responsible for information and data banks—will have to adjust their programs if they are to be effective in the changed setting.

Providing a nourishing climate for S&T, especially work that crosses traditional disciplinary boundaries, will be a challenge for all participants in the enterprise. Universities will need to craft means for strong multidisciplinary research without impairing the quality of the research enterprise. They will also need to protect their ability to conduct long-term research that may be of high risk even as pressure for goal-oriented research grows. Given that central corporate research is likely to continue to decline and that intense competitive pressures will continue to build for technology-based industries, the government must foster long-term and fundamental research across the broad frontiers of science and technology. Industry for its part must continue to recognize the critical value of academic research in enriching the base of fundamental knowledge on which it depends and the role of the government in supplying long-term support, or "patient capital." It will also be important for graduate education to be designed so that it equips new researchers to embark on satisfying careers and gives them the ability to respond quickly to new research opportunities, especially in interdisciplinary settings.[1]

---

[1]These points have been made in many previous articles and reports. Pressures on and changes in universities have been the subject of several collections of articles, including Ronald G. Ehrenberg, ed., *The American University: National Treasure or Endangered Species*. Ithaca, N.Y.: Cornell Uni-

Social beliefs and values will affect S&T, both fostering and hindering its adoption and range of uses. The onset of the environmental movement some 30 years ago affected both the directions and uses of new knowledge—for example, its incorporation into green industrial processes and products. That history may be echoed in how the new powers enabled by biological and medical advances are applied. Clinical therapies reliant on molecular and cellular manipulation will continue to be examined extremely closely and held to a very high standard of efficacy. The application of genetically modified organisms in agriculture may encounter much resistance. From a different perspective, public pressure may rise for technologies that give people more control over their own medical information and care.

- Pull factors driven by national needs and consumer demands also play a large role in shaping science and technology.

The United States and the rest of world face a number of problems that S&T could help resolve or mitigate. These problems create needs such as better technologies for aging and increasingly urban populations, economic development that is sustainable and internationally competitive, new sources of energy and improved energy efficiency, environmental protection and restoration, ways of coping with global climate change, education for a knowledge-based economy and a technically literate polity, and prevention and treatment of infectious and chronic diseases. The events of September 11 and the subsequent mailings of anthrax spores to public figures have sharply heightened the demand for better ways to prevent and recover from acts of terrorism, which will no doubt substantially affect the direction of research and technology development in a broad range of fields.

Individual consumer preferences and needs will also affect the demand for research and technology development. To take an obvious but important example, Moore's law—the doubling of computer processor power every 18 months—may be technically achievable, but it will not be achieved if consumers decide they do not need the increased capacity to accomplish what they want with computers and the Internet.

---

versity Press, 1997; Roger G. Noll, ed., *Challenges to Research Universities*. Washington, D.C.: Brookings, 1998; and William G. Bowen and Harold T. Shapiro, eds., *Universities and Their Leadership*. Princeton, N.J.: Princeton University Press, 1998. For changes under way in industrial research, see Charles F. Larson, "Industrial R&D in 2008," *Research•Technology•Management*, November-December 1998. The need to broaden the education and training of researchers is the subject of National Research Council, Committee on Science, Engineering, and Public Policy, *Reshaping the Graduate Education of Scientists and Engineers*. Washington, D.C.: National Academy Press, 1995.

- Although it is possible to discuss trends in science and technology and the factors that affect them, uncertainty about the future remains very high. In addition to the uncertainty surrounding the nature and timing of research advances and technological innovations, a complex set of contextual and demand factors, whose outcome we cannot predict, also affects trends.

The uncertainty about the future directions of S&T is not a failure of effort to understand but is inherent in the process of innovation. But even if it is impossible to make reliable forecasts of technological futures, especially in the dynamic fields of most interest, it certainly is possible to frame plans that are adaptive in their design and thus robust against a range of alternatives.

For example, policies should ensure that there is support for long-range research that the private sector cannot justify funding. Supporting basic research in the broad categories described in this report would offer a promising approach without requiring specific predictions of what breakthroughs will actually occur. There should also be adequate support of the infrastructure for innovation, including facilities and equipment, standards, and undergraduate and graduate education. Furthermore, since breakthroughs are likely to involve more than one basic field, efforts to promote and facilitate multidisciplinary research and effective interchange between disciplines of information about research advances can be quite valuable.[2]

Over and above these efforts, the recognition that citizens of democratic societies will more and more both affect and participate in decisions related to technological development suggests the strong need to promote technical literacy and to improve communication between the S&T community and the public at large. Industry, researchers, and public institutions would have to be involved in this work, which should be viewed as an integral part of the nation's research and development effort.[3]

---

[2]The need for sustained federal support of basic research and an adequate research infrastructure, and of interdisciplinary activities, is analyzed in a number of recent reports. The necessity of a federal role in supporting long-term, high-risk R&D is pointed out in National Science Board, *Federal Research Resources: A Process For Setting Priorities*. NSB 01-156. Arlington, Va.: National Science Foundation, 2001. For a statement of the need to ensure that the federal research portfolio matches national priorities and, in the face of uncertainty, also invests broadly in all areas of research, see National Research Council, Board on Science, Technology, and Economic Policy, *Trends in Federal Support of Research and Graduate Education*. Washington, D.C.: National Academy Press, 2001, and National Research Council, Committee on Science, Engineering, and Public Policy, *Observations on the President's Fiscal Year 2002 Federal Science and Technology Budget*. Washington, D.C.: National Academy Press, 2001. Another report analyzes the benefits of multidisciplinary research in the biology and information fields and recommends policies to foster more of it. See National Research Council, Board on Science, Technology, and Economic Policy, *Capitalizing on New Needs and New Opportunities: Government-Industry Partnerships in Biotechnology and Information Technologies*. Washington, D.C.: National Academy Press, 2001.

[3]National Science Board, *Communicating Science and Technology in the Public Interest*. NSF 00-99. Arlington, Va.: National Science Foundation, 2000.

Finally, if research and applications are to become more linked, the institutional sectors of the innovation system—industry, academia, nonprofit research institutes, and government—would have to develop and expand mechanisms for interaction and for coordinating their activities more closely. The challenge will be to accomplish this without losing sight of or compromising their respective strengths and unique roles. True synergy requires proper linkages between these various institutions rather than a blurring of their distinct purposes, and U.S. society is most likely to benefit from technological advances when each of the institutions involved in the innovation system functions effectively and in balance with the others.

Appendixes

A

Biographical Sketches of Committee Members

Kenneth H. Keller, *Chair*, is director of the Center for Science, Technology, and Public Affairs and professor of chemical engineering and materials science, both at the University of Minnesota. His current research, writing, and teaching deal with a wide range of public policy issues arising out of and related to the influence of science and technology on international affairs, the development and adoption of high-technology approaches to health care, and the role of American institutions of higher education in research and development. He was a senior fellow for science and technology at the Council on Foreign Relations 1990-1996 and serves or has served on the boards of RAND's Institute for Education & Training and the Science Museum of Minnesota. He earned his M.S. and Ph.D. (chemical engineering) from Johns Hopkins University.

Milton Chang is with iNCUBiC, LLC. Before that, he was president of New Focus, Inc., which supplies photonics tools for laser applications and which he founded in 1990. He has incubated or is actively incubating 16 companies, including Uniphase and Gadzoox Networks. He currently serves on the boards of companies he has incubated, which span a broad spectrum of industries: Agility Communications, Arcturus Engineering, YesVideo.com, Euphonix, Iridex, Gadzoox Networks, New Focus, LightConnect, Lightwave Electronics, and Optical MicroMachines. He is active in the photonics community and has received a number of awards from professional societies. He writes monthly business columns for *Laser Focus World* and *Photonics Spectra*. From 1996 to 1998, Dr. Chang served on the Visiting Committee for Advanced Technology of the National Institute of Standards and Technology. Dr. Chang holds a B.S. in electrical engineering from the University of Illinois and an M.S. and a Ph.D. in electrical engineering from the California Institute of Technology.

William E. Coyne has been the senior scientist among 7000 researchers in 3M laboratories around the world, overseeing 30 different technology platforms. He joined 3M as a research chemist in 1968 and retired last year as senior vice president, R&D, and as a member of the company's senior executive team. During his tenure as the senior architect of 3M's R&D, the company invested more than $1 billion a year in research and significantly raised its new-product targets. 3M recently set a record of $1.4 billion in first-year new product sales and also received the coveted U.S. Medal of Technology from President Clinton. Dr. Coyne also sponsored the 3M Technical Forum, an organization of thousands of 3M researchers who meet periodically to share research, technology, and ideas. He earned his Ph.D. in organic chemistry from the University of Virginia and B.S. and M.S. degrees in pharmacy and pharmaceutical chemistry from the University of Toronto.

James W. Dally (NAE) is Glen L. Martin Institute Professor of Engineering emeritus at the University of Maryland, College Park. Dr. Dally, a mechanical engineer, has had a distinguished career in industry, government, and academia. His former positions include dean of the College of Engineering at the University of Rhode Island, senior research engineer at Armour Research Foundation, professor and assistant director of research at the Illinois Institute of Technology, and senior engineer at IBM. He is author or coauthor of six books, including engineering textbooks on experimental stress analysis, engineering design, instrumentation, and the packaging of electronic systems and has published approximately 200 research papers. He earned his B.S. and M.S. from Carnegie Institute of Technology and his Ph.D. from the Illinois Institute of Technology.

Charles P. DeLisi is Metcalf Professor of Science and Engineering at Boston University (BU) and also served as dean of the College of Engineering from 1990 to 2000. Prior to moving to BU, he was professor and chair of biomathematical sciences at Mount Sinai Medical School (1987-1999), director of the Department of Energy's Health and Environmental Research Programs (1985-1987), section chief at NIH (1975-1985), and Theoretical Division Staff Scientist at Los Alamos (1972-1975). He has authored or coauthored more than 200 scientific papers and has five patents. Dr. DeLisi is a fellow of the American Association for the Advancement of Science and the American Institute of Medical and Biological Engineers and is the recipient of numerous awards, including the Smithsonian Platinum Technology Laureate, the Department of Energy's Distinguished Associate Award from Secretary Richardson and the Presidential Citizen's Medal, awarded to him in January 2000 by President Clinton for initiating the Human Genome Project. His B.S. (from City College of New York) and his Ph.D. (from New York University) are both in physics; his postdoctoral training (at Yale) was in chemistry.

C. William Gear (NAE) is former president of NEC Research Institute. Prior to joining NEC in 1990, he was head of the Department of Computer Science and professor of computer science and applied mathematics at the University of Illinois at Urbana-Champaign. His research expertise is in numerical analysis and computational software. He is an IEEE fellow, served as president of the Society for Industrial and Applied Mathematics (SIAM), and was a recipient of the Association for Computing Machinery (ACM) SIGNUM George E. Forsythe Memorial Award as well as Fulbright and Johnson Foundation fellowships. He earned his B.A. from Cambridge University and his M.S. and Ph.D. (mathematics) from the University of Illinois.

Roy Levin is now with Microsoft Corporation, where he is the director of its Silicon Valley Research Lab. Before that, he was director of the System Research Center at Compaq, where he worked since the company's founding by Digital Equipment Corporation in 1984. Prior to that he was a principal scientist at Xerox's Palo Alto Research Center (PARC), where he was a principal developer of Cedar, an integrated programming environment, and Grapevine, one of the earliest e-mail systems. Dr. Levin earned his B.S. in mathematics at Yale University and his Ph.D. in computer science at Carnegie Mellon University. He is a member of the Association for Computing Machinery and author or coauthor of approximately 20 technical papers, books, and patents.

Richard L. Popp is professor of medicine at Stanford University School of Medicine, where he has taught since 1971. Dr. Popp has directed the Stanford-Israeli Visiting Professors Program since 1984, is a member of the New York Academy of Sciences and the California Academy of Medicine, and is active in a number of professional associations. He was honored by the American Society of Echocardiography with the creation of its Richard L. Popp Master Teacher Award, to be presented annually. He earned his B.A. and M.D. degrees from Johns Hopkins University and its School of Medicine.

Nathan Rosenberg is Fairleigh S. Dickinson, Jr., Professor of Public Policy in the Economics Department at Stanford University. His research interests encompass the economics of technological change, the economic role of science, and economic history and development, with current interests including the role of scientific knowledge in influencing the rate and direction of technological change, determinants of technological change in the chemical and medical sectors, and the economic performance of high-tech industries. Recent books include *Technology and the Pursuit of Economic Growth*, 1989 (with David Mowery); *Exploring the Black Box*, 1994; and *The Emergence of Economic Ideas*, 1994. He earned his Ph.D. from the University of Wisconsin and his A.B. from Rutgers University.

Thomas A. Saponas is a senior vice president and CTO of Agilent Technologies as well as director of Agilent Laboratories. His responsibilities include developing the company's long-term technology strategy and overseeing the alignment of the company's objectives with its centralized R&D activities. He has more than 27 years of experience in electrical engineering refined over the course of his career with Hewlett-Packard Company, where he began in 1972 as a design engineer in the company's Automatic Measurement Division and went on to become vice president and general manager of the Electronic Instruments Group. In 1986, Mr. Saponas was selected to serve as a White House fellow and served as special assistant to the Secretary of the Navy for a year, on leave from Hewlett-Packard. He earned a B.S. in electrical engineering and computer science and an M.S. in electrical engineering from the University of Colorado.

B

Workshop Agenda

Friday, July 20, 2001

3:00 pm   Welcome. Introductions. Discussion of agenda, assignments, brief comments from the NIST acting director.

4:00   Paper on trends in science and technology, presented by the author, Patrick Young. Brief discussion.

4:30   Breakout groups session 1.

6:30   Adjourn. Reception with local corporate leaders, museum officials, and others.

7:30   Dinner at museum. Museum tour to follow.

Saturday, July 21

8:30 am   Convene in plenary session.

8:45   Paper on trends in industrial management and organization, presented by Stephen Merrill. Brief discussion.

9:15   Breakout groups session 2.

Noon   Lunch.

1:00 pm   Paper on trends in the economy and industrial strength, presented by Norman Metzger. Brief discussion.

1:30   Breakout groups session 3.

5:00   Adjourn. Individual dinner arrangements. Breakout group chairs prepare reports to be given next morning.

Sunday, July 22

8:30 am	Convene. Brief amplification by breakout groups of template items.
9:15	Plenary discussion.
10:30	Break.
10:45	Continue plenary discussion.
Noon	Workshop adjourns.
12:30 pm	Working lunch of committee to outline major themes in workshop report.
2:30	Committee adjourns.

C

Workshop Participants

INVITED PARTICIPANTS

David H. Auston, former President, Case Western Reserve University
Thomas M. Baer, President and CEO, Arcturus Engineering
David C. Bonner, Global Director, Polymer Technology Center, Rohm and Haas Company
Wilmer R. Bottoms, CEO and Founder, Third Millenium Test Solutions
Bruce G. Buchanan, Professor of Computer Science, University of Pittsburgh
Robert Buderi, Editor at Large, *Technology Review*
Charles R. Cantor, Chief Scientific Officer, Sequenom, Inc.
Linda A. Capuano, Vice President e-Business Integration, Honeywell Engines & Systems, Honeywell, Inc.
Robert P. Caren, Vice President (retired), Sciences and Engineering, Lockheed Corporation
James Flanagan, Vice President, Information Processing Research, Avaya, Inc.
L. Charles Hebel, Manager of Technology Evaluation (retired), Xerox Palo Alto Research Center
Robert J. Hermann, Senior Partner, Global Technology Partners, LLC
Christopher T. Hill, Vice Provost for Research and Professor of Public Policy and Technology, George Mason University
Deborah A. Joseph, Associate Professor of Computer Sciences and Mathematics, University of Wisconsin at Madison
Donald B. Keck, Vice President and Executive Director, Office of Research, Corning, Inc.

Paul Mankiewich, Chief Architect and CTO, Wireless Network Group, Lucent Technologies
R. Byron Pipes, President (retired), Rensselaer Polytechnic Institute
John T. Preston, President and CEO, Atomic Ordered Materials
Syed Z. Shariq, RGK Foundation Scholar and Leader, RGK Program on the Knowledge Economy (Knexus), Stanford University
Kathleen C. Taylor, Director, Materials and Processes Laboratory, General Motors Corporation
Hal R. Varian, Dean and Professor, School of Information Management Systems, University of California at Berkeley
Dennis A. Yao, Joel and Anne Ehrenkranz Family Term Associate Professor, Wharton School of Business, University of Pennsylvania

COMMITTEE MEMBERS

Kenneth H. Keller, University of Minnesota, Chair
Milton Chang, iNCUBiC, LLC
William E. Coyne, 3M Corporation
James W. Dally, University of Maryland at College Park
Charles P. DeLisi, Boston University
C. William Gear, NEC Research Institute, Inc.
Roy Levin, Microsoft Corporation
Richard L. Popp, Stanford University School of Medicine
Nathan Rosenberg, Stanford University
Thomas A. Saponas, Agilent Technologies, Inc.

NIST OBSERVERS

William E. Anderson, Director, Electronics and Electrical Engineering Laboratory
Howard M. Bloom, Acting Director, Manufacturing Engineering Laboratory
Karen H. Brown, Acting Director, NIST
Kevin M. Carr, Director, Manufacturing Extension Partnership Program
Michael P. Casassa, Acting Director, Program Office
Paul N. Doremus, Strategic Planning Analyst
Katharine Blodgett Gebbie, Director, Physics Laboratory
Harry S. Hertz, Director, Baldrige National Quality Program
William O. Mehuron, Director, Information Technology Laboratory
Hratch G. Semerjian, Director, Chemical Science and Technology Laboratory
Leslie E. Smith, Director, Materials Science and Engineering Laboratory
Jack E. Snell, Director, Building and Fire Research Laboratory
James A. St. Pierre, Director, Industrial Liaison Office

NRC STAFF AND COMMISSIONED AUTHORS

Maria Jones
Michael McGeary
Stephen A. Merrill
Norman Metzger
Marsha Riebe (University of Minnesota)
Patrick Young
Dorothy Zolandz

D

Workshop Summary

INTRODUCTION

The workshop held by the Committee on Future Environments for NIST on July 20-22, 2001, in St. Paul, Minnesota, was attended by committee members, invited experts from various sectors of industry and academia (see Appendix C for the list of participants), and staff. In addition, the acting director and a dozen other leaders of NIST were present as observers. All participants were provided with copies of the commissioned papers (Appendixes E through I). Three of the commissioned papers (those by Young, Bean, and Finneran) were presented during the workshop. A large part of the workshop was devoted to three half-day sessions of breakout groups. These sessions were organized along the three themes selected by the committee: "push," "contextual," and "pull" factors. The push factors session gathered together judgments on possible futures for a set of scientific and technical fields. The pull factors session focused on societal demand factors and the economic, social, environmental, and political needs and sensitivities that would promote or inhibit research and development in certain areas of science and technology as well as innovations based on that research. The contextual factors session considered issues such as changes in the organization and support of R&D in both the public and private sectors, educational goals of students and methods of delivering education, and patterns of investment by the private sector, all of which might be expected to change the process by which ideas move from research to product. While obviously this organizational principle is somewhat arbitrary, the committee nevertheless found the principle a powerful one, given the task before it.

The workshop proceedings are organized around the topics considered by the breakout groups at the three sessions (see Appendix B for the workshop agenda). In the first session, each breakout group was asked to consider push factors parsed into science and technology trends in three fields—one group considered the biological sciences; another considered materials science and technology; and the third considered computers, communication, and information technology. At the second and third sessions, every group dealt with the same topics and questions. In the second session, which focused on contextual factors, the three breakout group topics were the organization of research, investment patterns, and public policy issues, respectively. In the third session, which focused on push factors, the groups discussed social and cultural trends and attitudes; economic considerations; and the international context.

The chair of each of the three groups was given a template—a form on which to summarize the results of their three sessions—for cross-comparison in a plenary discussion that was held on the last day of the workshop. In addition, on that last day, the template-ordered results were displayed and distributed, and the entire morning was devoted to brief reports from the breakout groups, followed by plenary discussion.

The committee is the first to recognize that analyzing trends in science and technology by dividing them into push, pull, and contextual factors is arbitrary when discussing any particular technological development. It is attempted here, at the risk of overlap and redundancy, to avoid the trap of technological determinism, that is, assuming that what is technologically possible will be used as soon as it can be developed. The rate of innovation—that is, the speed at which scientific advances are made and turn into economically viable technologies—also depends on the infrastructure for innovation and, perhaps more important, on social and economic factors. The innovation infrastructure includes the supply of human capital afforded by a nation's education and training systems and the funding of research and technology development by industry, venture capitalists, nonprofits and foundations, and government. Social and economic forces accelerating or impeding technology are legion. The committee's approach recognizes that social and economic factors play a large role in the fate of any technology, and it underlines the importance of the innovation infrastructure in enabling economic growth and increased productivity.

BREAKOUT SESSION 1: PUSH FACTORS

The purpose of this session was to identify and discuss the scientific advances and new technological developments that will create the potential for new applications in the next decade. Each subgroup was asked to consider a different set of questions, focused in a particular area of science, although they were all encouraged to address new developments arising at the intersections of the areas.

Group on the Biological Sciences

Molecular and Cellular Biology

1. It is possible that for each of the major diseases a relatively small number of genes—probably hundreds rather than thousands—will be identified over the next few years, selected, and studied extensively. Large-scale screening of genes will be less practiced in the future.
2. The large-scale screening of proteins is not likely to be an effective approach, because analyzing amino acids (proteomics) is more difficult than analyzing nucleic acids (genomics). Some of the most important protein structures are still unknown.
3. The technology for protein analysis is improving (mass spectrometry and other technologies are being used). The advance in protein analysis is hindered by the difficulty of crystallizing materials for analysis, but efforts will be made to improve that process.
4. There is a sequence of tasks involved in understanding the role of genetics in human health and disease—identifying the gene, then the protein, the protein structure, reaction pathways, cell biology, organ function, and so on. In next 10 years, work will concentrate on the gene, the protein, and the protein structure; the remaining steps pose more difficult challenges.
5. Genetic, immune, and small-molecule therapies are possible applications. Custom drug design is possible, but companies are concerned about minimizing drug side effects as well as maximizing efficacy.
6. Multigene traits are more important in agriculture than in human health, and they pose a much more complex computational problem.

Medical Devices and Instrumentation

1. There will be important advances in tissue engineering, based on better understanding of synthetic surfaces and how to characterize and modify them, and on improved understanding of cell membrane structure and function. Interfacial science—that is, the study of the interfaces between biological and man-made materials—will drive and be driven by tissue engineering.
2. It is now possible to generate certain cell populations—beta cells, liver cells—that are two-dimensional tissues (epithelium, endothelium, vasculature). There has been progress in regenerating nerve cells. Synthetic materials can be used to guide the morphology of cell growth and proliferation, develop hybrid organs, and protect cells against rejection. But the ability to grow whole natural organs is unlikely to be achieved in the next decade.
3. The use of MEMS for in situ intelligent sensors is growing in importance.

Microsurgical instruments can be operated remotely, which allows less-invasive surgery. Robotic instruments are already used in neurosurgery and eye surgery.
4. More and more site-specific imaging markers are becoming available, which is leading to advances in imaging.
5. There are new imaging techniques by which to study brain metabolism, which is enabling better interpretation of functional MRI images.

Cross-Cutting Applications

1. As classical functional taxonomy shifts to the genetic classification of organisms, it will be important to identify the traits to focus on, so as to reduce to a manageable amount the information that must be sifted through.
2. Advances in genomic data on plants will improve the analysis of inadvertent sequence changes during the genetic modification of organisms (and also during traditional plant breeding).
3. Biological computers based on DNA will be possible, but they will be curiosities rather than practical devices because of the limited range of problems they can solve.
4. There will be greater use of the microorganisms known as extremophiles for environmental cleanup (oil spills, etc.). Such organisms can function in conditions we have not been able to manage before, such as high heat or pressure or in the absence of oxygen, and they are beginning to be used as catalysts in fine chemical production.
5. It may become possible to use biotechnology for high-efficiency energy applications.

Group on Materials Science and Technology

Directions in Materials Science

1. Atom-by-atom customization of commercial products will come about in the next decade, enabling the surface properties of materials to be manipulated. This in turn will impart functionality to the materials, allowing them to be used in many different applications.
2. There will be further development of photonic crystals and their applications in communications.
3. There will be new materials requirements for electronics: low-K dielectrics with acceptable mechanical properties and high-K dielectrics with low trap and interface states.
4. Nanotubes are suitable for a number of specific applications. However, the ability to process and prepare them in commercially useful forms and

quantities, and at acceptable cost, has been a substantial barrier but will probably be overcome during the next decade.
5. Because of the small size and optical properties of quantum dots, the technology for producing and using them will have potential application in computing, optical communications, and biological and chemical sensing.
6. Instrumentation and techniques to identify, measure and analyze, and standardize exist for bulk materials but not for nanomaterials. It is expected that they will be developed during the next 10 years.

Advances in Materials-Dependent Applications

1. The efficiency of fuel cells has increased greatly over the past decade. They may one day replace batteries for consumer products, but the fuel, hydrogen, is difficult to carry in an electric automobile. Either an alternative fuel must be found or an infrastructure developed to supply hydrogen. The first important application for fuel cells and other technologies in the next 10 years will probably be in local power generation for the power grid.
2. Light-emitting diodes (LEDs) are a promising low-energy source of white light and energy and electric power, if the problems of producing low-cost blue LEDs or organic LEDs (thin films) can be solved.
3. There is a need for catalysts to produce lower-cost, more-efficient materials. Nanostructures might be used as catalyst carriers.
4. There will be advances in the development and use of bioremediation materials, whether biological or other materials.

Integrated Technologies

1. Thermionic cooling will be developed as an alternative to refrigerants, which harm the ozone layer.
2. MEMS will be applied for new methods of optical switching and self-replicating/self-organizing/self-assembling.
3. The technology for organizing quantum dots into useful systems, which may be developed during the next 10 or 20 years, will have important applications in archival data storage devices.
4. MEMS devices could be developed for microfactories that can generate and deliver power on a local basis without needing more transmission lines.
5. MEMS will be used for cell sorting.
6. Integrated systems of sensors and communications will be developed for medical care and other applications. MEMS, with their sensor capacities

and delivery capabilities, may succeed in making artificial organs within the next 10 years.
7. New materials will be developed for the separation sciences.

Group on Computers, Communication, and Information Technology

Advances in Computers and Communication

1. Component-level advances will continue through the decade.
2. Storage will become sufficiently cheap to permit continuous recording of large amounts of visual and physiological parameters, and reductions in the size of storage devices will enable complete personal information to always be in each person's physical possession.
3. Advances in computation and other components will allow very smart, very small, and very cheap sensors that can be widely distributed and communicate easily with networks.
4. Simulation and modeling will continue to displace physical experimentation.

Advances in the Application of Information Technology

1. The complexities of large systems will limit the rate at which the substantial increases in computational and networking power can be applied.
2. There is a long history of innovative applications of artificial intelligence, and advances in sensors that are intelligent and able to integrate information will greatly improve a system's ability to interpret data in context. Speech recognition has been improving tremendously, but speech understanding under natural conditions requires a degree of context that is still not possible. Visual recognition will still be difficult, even with 100-fold CPU speed advances expected during the next decade, although existing capabilities are used routinely by the military—for example, for target recognition by moving aircraft. Advances in personalized interfaces discussed above would help, because each person could provide the computer with personal data that would ease the recognition and some of the understanding problems.
3. Systems are gradually improving at performing both self-diagnosis and self-correction, which will help manage the next level of system complexity.

Human–Machine Interface Issues

1. Interface issues will dominate, and limit, the adoption of most new applications. Interfaces have to be natural and easy to learn. An example of a

successful interface is the cellular telephone, which the consumer can easily use despite the sophisticated computation and networking involved. Unless there are major advances in interfaces, there will be limited commercial demand for more increased complexity or further performance breakthroughs in computers and networks.
2. Primitive telepresence will be available within 10 years, but it will take a two-orders-of-magnitude improvement in computation and communication for videoconferencing to begin to substitute seriously for face-to-face communications.
3. Personalized interfaces that recognize individuals and their habits will enable less sophisticated users to tap into high-powered computers and networks.
4. Better technology for secure data transmission, storage, and authentication will be developed, but privacy issues cannot be solved just by better technologies. They involve trade-offs and choices. For example, a person carrying his or her complete medical record may want total control over access to the record, but what if that person arrives in an emergency room unconscious?

Plenary Discussion of Push Factors

The first observation was that separating technology push issues from societal pull issues conceptually is very difficult to achieve in practice. Those entities developing technologies are well aware of the economic, social, political, and legal opportunities and barriers. Trends in industrial research, especially the development of closer linkages between commercialization and research, are imposing an even greater consideration of pull factors in decisions on research directions and technology development. The subgroup on push (supply) factors in computational science and information technology devoted most of its time to the need for improvements in machine–human interfaces in order for computer and information technology to develop further. If technologists do not improve those interfaces, demand for more advanced technologies will be dampened by the reluctance of users to buy more complex products.

There was a discussion of the potential for advances in artificial intelligence (AI). One participant argued that there has been a long history of innovative applications of AI, and many Fortune 500 companies are using it routinely. The development of better sensor devices will enable more intelligent integration of information and improve the capacity of systems to interpret data in context. Systems for visual and speech recognition are in routine use, for example, by the military in target recognition. Another participant agreed that AI has achieved much but said there was a long way to go to achieve speech understanding in other than controlled contexts.

The position that only a few hundred genes at most will be involved in studies of health and longevity was questioned. Recent understanding from sequencing the human genome has moved past the one-gene, one-disease paradigm, and gene chip technology may show that one disease is associated with the expression of many genes. One response was that it looks as though only a few primary genes are involved in each major disease, although there may be many downstream modifiers. Gene expression chips allow identification of many genes involved in a disease, but the chips show the effects of the disease, which can be very complex, rather than its causes, which can be very simple.

It was observed that the issues addressed by the groups did not include trends in manufacturing, which is still an important although declining set of activities. This turned the discussion to the reports from the second breakout session, which focused on, among other things, shifts in the organization of research, development, and innovation.

BREAKOUT SESSION 2: FINDINGS ON CONTEXTUAL FACTORS

Organization of Research

1. Universities are important in fostering industrial innovation and speeding the commercialization of innovation, but needed changes in their organization and practices are taking place only slowly:

   - Traditional academic organization by disciplinary departments does not facilitate the interdisciplinary training that industry wants, although universities are trying to respond to the demand for such training.
   - The incentive system in academia is changing. Students are now more interested in patents than papers, and the best graduates are going into industry (especially small companies), not academia, which may make it difficult to maintain the quality of university faculties.
   - The review and reward system should be revised to promote interdisciplinary research and graduate training.
   - Successful communication across disciplinary boundaries must be ensured. Vocabulary differences between disciplines must be bridged.
   - Higher education should provide more effectively for the technical agility of the workforce, and education and experience need to pay more attention to entrepreneurial skills and teamwork.
   - Scientists and engineers need to be taught to be integrators across disciplines, including business, manufacturing, patents, etc. This will require strong continuing education programs.
   - Large global universities will emerge, and private universities will have to specialize to become global centers of excellence. MIT, for example, is moving its Media Labs into India and Europe.

2. The organization of research is changing and will continue to change in industry. The trend is toward decentralization within companies, greater outsourcing, and partnerships with other companies and research institutions.

- Corporate central research organizations are in decline. Universities could become the suppliers of just-in-time knowledge. Venture-capital-funded enterprises and independent labs like SRI and Sarnoff may also replace internal corporate research operations.
- There will be more cost sharing through jointly owned central research labs. For example, Rockwell Science Center is now owned jointly by two separate companies, Rockwell Collins and Rockwell Automation, and HRL Laboratories, formerly Hughes Research Laboratories, is owned by Boeing, General Motors, and Raytheon.
- Roadmap activity in some research areas is not adequate and would help focus R&D on needs. This should be an industry activity, but government could catalyze roadmapping and make some changes to avoid inhibiting collaboration. This may be difficult for some industries, for example, the pharmaceutical industry.
- Mobility of people and ideas is a competitive advantage for industry as a whole, although out of self-interest, an individual company will typically resist mobility at the margin.
- Technology development will be significant in new companies and will be marketed most effectively by larger companies with the necessary infrastructure. This is an underpinning of a successful mergers and acquisitions activity.
- Larger companies will need to organize project teams internally (and perhaps externally) to foster rapid innovation and accelerate the integration of technologies.
- Alliances between companies can facilitate development and marketing and will be an important activity in the future. Alliances speed time to market and facilitate specialization and learning, but they also pose serious organizational interface challenges for and could cause conflict between longer-term strategic goals. Unsuccessful alliances are more common than successful alliances.
- Geographically distributed research is already a reality for information sciences, but it is more difficult to achieve when physical laboratory facilities are involved.
- Merger and acquisition activity among large companies generally decreases investment in R&D—companies merge, there is overlap, and the number of people involved in R&D is cut—and does not foster innovation.

- The benefits of globalization, including 24/7 project activities and greater diversity of thinking, are increasingly important.

3. Interactions and relationships between academia and industry are becoming closer, more frequent, and more complex, and the roles of faculty and students in industrial activities and their implications are creating issues that must be worked out.

    - The university-industry interface varies by field. It is more effective in the life sciences and is more likely to continue there than in the physical sciences. The difference stems in part from the looser connection between research and application in the physical science areas and the large capital investment that industry itself is making, which lessens the need for university relationships (universities are still important for supplying highly trained physical scientists and engineers).
    - The high quality of industrial research is increasingly attracting faculty to take their sabbaticals in industry or otherwise move between the private sector and universities. This trend involves trade-offs as faculty allocate time between their start-ups and their students.
    - University faculty have become involved in company start-ups as entrepreneurs. There is increased collaboration between universities and industry, including equity participation in start-ups based on university technology and resources. Conflicts of interest and/or the appearance of conflicts of interest must be addressed as collaborations and entrepreneurship increase. Public disclosure of all information will be a key element in addressing this issue.

4. The organization of federal research agencies around missions and diseases is not conducive to support of basic research. For example, the orientation of the NIH toward specific diseases and disease hypotheses does not encourage the kind of multidisciplinary collaboration on basic phenomena that leads to fundamental advances even though it is not aimed at curing a particular disease.
5. Some other trends in the organization of research are these:

    - Financial rewards for researchers vary greatly from sector to sector—government, academia, large corporations, and venture capital start-ups.
    - There is no silver bullet in organization structure. Whatever the structure, it must include business knowledge and not inhibit innovation while keeping that innovation focused on the commercial objectives.
    - The free market is a major asset in providing rewards and recognition and handles this well. Some are concerned, however, that the system

negatively affects the advancement of some employees—for example, women with families who cannot put in long hours at critical phases in their careers.

Investment Patterns

1. Innovation is capitalized quite well by both public and private capital under the patterns that exist today. However, environmental, political, and other issues may change this in the future, and new accounting rules may affect investment patterns.
2. The venture capital model is creating multiple points of technology innovation by funding many small companies, especially in the biotechnology and information technology areas, that are pursuing parallel research.
3. Funding models (e.g., small, short-term grants) conflict with the creation of centers of excellence, which require stable, long-term support.
4. Support of basic research is in question and needs to be addressed in both government and industry. Such research takes place largely in universities, and sources of funding need to be identified. But public funding of research may increase, owing to public anxiety about terrorism.
5. Short-range accountability for results is hindering long-term research as well as the development of infrastructure necessary for greater innovation in the longer run:

   - Innovation in industry is constrained by a need to have good quarterly results, especially among companies that go public early. Long-term research is less constrained by short-term financial expectations in private companies.
   - In some particular areas, metrology and standards require more investment. Deficiencies exist today in standards, materials properties, and metrology that would facilitate the insertion of innovation into the economy.
   - Infrastructure needs that are not adequately supported by the profit motive may require government resources.

6. Peer review has become more conservative as the number of applications increases relative to the amount of funding. Mechanisms to encourage risk taking need to be developed.
7. Despite globalization of R&D and innovation, regional clustering of innovation remains important because it concentrates resources and satisfies the need for critical mass. Close connection to customers and their needs also supports localization despite global ownership.
8. There are more and more opportunities for private and individual investment in education and continuing education:

- Information technology developments will cause major changes in education. Formal education and continuing education will become less important as it becomes possible for individuals to solve problems from information they can gather rather than from knowledge they possess.
- There is a rapidly growing for-profit education sector, including credentialing, e.g., M.B.A.'s.
- Effective techniques to invest in educational needs should be identified.

Public Policy Issues

1. Workforce needs of the future will not be met by the educational system without some changes. The gap is widening between traditional K-12 education and a technology-driven society. The problem needs to be addressed at the primary and secondary school levels.
2. A large and growing percentage of graduate students and Ph.D.'s is foreign, which may become a problem if the supply drops off, but the trend does promote communication across national borders as some of these people return to their own countries.
3. The U.S. intellectual property regime—including patent and copyright laws—is currently a constraint, as innovation accelerates:

   - Although intellectual property issues may be on the path to resolution, there is still a problem in newer fields.
   - New waves of innovation usually result in patent wars. The costs of litigation must be minimized.
   - U.S. patenting policy on gene sequences needs revision if multiple gene applications are to be encouraged.
   - Patent law in new areas such as certain biotechnology areas may have several years of uncertainty that may inhibit investment in innovation.
   - U.S. copyright policy and law also currently inhibit progress.
   - Differences between U.S. patent and copyright policy and patent and copyright policy of our trading partners need to be resolved.

4. There is a growing need for standardization. For example, standards are needed in genomics/proteomics (judging sequence accuracy, standardizing software, and so on), which will not be met by the National Institutes of Health or the Food and Drug Administration. Also, there are data format issues (longevity and data migration) that require formal standards. However, while physical standards are always useful, they sometimes serve to inhibit trade and other times to promote it. Defining standards so that they promote it will be a major challenge.

5. The security and validity of online information are becoming an important issue.
6. There are some other impediments as well:

- Provisions in laws and regulations that inhibit adoption of technology without reason (some of them are more than a hundred years old) should be reviewed and abolished or modified, as was done in Massachusetts.
- Regulation and taxation of Internet activity may inhibit progress.
- Foreign outsourcing of research raises transfer of technology issues.

Plenary Discussion of Contextual Factors

Some participants argued that the assessment of the patent and copyright systems was too negative. They said that while not perfect, the systems are generally working, although there may be problems in the fast-emerging area of biotechnology. Others restated the case that the patent system is not keeping up with changes in science and technology and is unduly hindering positive developments. There were a number of examples of problems with patents, including patenting too early and too broadly, which hinders development; unrealistically high expectations on the part of universities that licenses will become a significant source of revenue; the use of defensive or blocking patents; and the reliance on litigation to make policy because initial patents are too easy to obtain. History shows that periods of accelerated innovation have usually led to patent wars, which imposes large financial costs and hinders innovation, and public policy should try to minimize such conflicts and litigation costs. There are also substantial differences in the patent and copyright policies of the United States and its international trading partners.

It was argued that differences between public and private research universities are small and narrowing. State funding is declining as a percentage of public university budgets, and public universities are increasingly dependent on the same sources of support (i.e., federal R&D agencies, foundations, and industry) as private schools. Also, research is globalizing as is the delivery of education, another reason to have a global university. State universities already offer programs abroad. Another participant agreed that public and private research universities are going in the same direction, but state schools will lag because of state legislative pressure to increase the emphasis on teaching relative to research. Also, the corporate status of state universities is not the same; some are more independent from state control than others.

One participant pointed out that the groups did not address trends in the status or role of the DOE national laboratories. In ensuing discussion, it was observed that the national laboratories are good at pursuing complex, long-term, interdisciplinary projects of the kind that industry values. The example given was the extreme ultraviolet lithography project, which involves three national laboratories and a consortium of companies, including Intel. On the other hand, there are constraints on the national laboratories when it comes to joint activities with other research institutions. The national laboratories are relatively isolated from industrial involvement, more so than universities, whose professors are increasingly taking sabbaticals to work in industry and whose former students working in industry provide useful feedback. The national laboratories have appointed individuals and established advisory groups to pursue industrial commercialization and venture capital resources and have met with varying degrees of success, although it is perhaps too early to evaluate such initiatives. The national laboratories are also diverse in their missions and roles, which makes it hard to generalize.

There was an extended discussion of the trend toward outsourcing of R&D by industry and the implications for future innovation of reductions in central research units. Companies can fund R&D internally, outsource it, or acquire it. It is an expense that, like other expenses, is judged on its rate of return. Not all research is a core competency that a firm will want to maintain internally. Will innovation tend to take place in small companies and be acquired by large companies, and will that offset the downsizing of central research programs? Or will it be outsourced to small firms or overseas? Are time differences and geographical distance an important constraint on outsourcing overseas? Will increasing dependence on outsourcing lead to inadequate research bases in the United States when foreign research laboratories become competent enough not to need U.S. support anymore? Are start-up companies sources of innovation or are they spun off from big companies or a university with a great idea? Do the central laboratories that remain no longer see themselves as independent campuses but as an integral part of the business development process? Will universities or national laboratories be able to become a more direct source of industrial innovation? Will national laboratories that move into new mission areas compete unfairly with industry because they are federally funded, or are they too shackled by government rules and regulations? The consensus was that it is too early to tell how it will all turn out or to know the consequences of, say, decentralizing R&D or moving it to foreign locations.

BREAKOUT SESSION 3: FINDINGS ON PULL FACTORS

Social and Cultural Trends and Attitudes

1. Consumers worldwide increasingly determine products and technologies, but there are significant regional differences in acceptance of new technologies (e.g., nuclear power, genetically modified organisms and foods, and medical devices). U.S. firms need to be aware of these differences in conducting R&D as well as in marketing.

2. Environmental concerns and the need for cheap sources of energy will be a major driver of technical developments as a result of both economic and political pressures. The U.S. economy is built on the assumption of cheap energy. The price of energy cannot be sustained without significant new technical developments.
3. Dramatic and unforeseen events will also affect the adoption of technologies. For example, according to public opinion polls there, the energy crisis in California has increased public acceptance of additional power from conventional and even nuclear sources.
4. There will be growing concern about the impacts of emerging technologies on values important to citizens—for example, the impact of information technologies on privacy and biotechnology's ability to genetically modify humans, animals, and plants—although the real risks are sometimes misunderstood:

- There is often a difference between the public perception of a risk and the objective assessment of that risk. These differences can lead to overestimates of risk (as in the case of nuclear power) and underestimates (as in the case of large software programs). The public, including the media, has the unrealistic expectation that risk should be zero, and it must be educated to better understand risk.
- In the United States, there is now general acceptance of genetically engineered drugs, such as t-PA and insulin, but growing concerns about genetically modified organisms and food.
- Free markets tend to respond quickly to needs, and public attitudes in a democracy may be quickly modified in response to any perceived crisis. Transparency and real-time availability of information will have an increasing impact on the development of public perceptions and therefore political views.

5. Concerns about the impacts of technologies vary from group to group within the United States and between nations:

- There are likely to be sensationalized negative portrayals of new technology, leading people to unreasonably fear that they will lose their privacy and control over important parts of their lives.
- There is a growing disparity between groups that benefit from technological advances and those that do not, which may lead to backlash against some new technologies by those not benefiting. Increasing cultural diversity within the United States and in world markets for U.S. goods and services makes reactions to events unpredictable.

- Different generations of Internet users may have substantially different sensitivities to the privacy (and other) implications of new technologies.
- Globalization may cause the United States to alter its positions on research and technology developments in response to the attitudes of other governments and cultures.
- The degree to which new technologies are adopted will be impacted by the technical literacy of a society.

6. Popular views of technology may impede research in certain areas but promote it in other areas:

- A substantial proportion of the public objects to the conduct of stem cell research and the creation of genetically engineered organisms (impede).
- Social issues have caused some medicines to be removed from the market (impede).
- Nuclear energy development in the United States has been inhibited, forcing the pebble-bed reactor to be built in South Africa (impedes some energy technologies while promoting others).
- California regulations may delay the acceptance of improved diesel engines in the United States (impede).
- Environmental concerns drive new market opportunities such as the development of recyclable polymers (promote).
- Green manufacturing and sustainable development will be demanded (promote).

7. There is a need for honest brokers to gauge the risk for emerging technologies and help build public confidence:

- Society needs mechanisms to instill trust and defuse the threat of the unknown. The National Academies have the potential to assist in this area by conducting reviews of technology impacts, but additional approaches to creating confidence in new technologies will be required.
- The science and technology communities have done things to undermine public trust. They need to learn what they must do to generate and maintain trust. This will require the fair communication of diverse viewpoints.

8. Human–machine interfaces need to improve generally to increase the use of, and therefore the demand for, high-tech products and services.

9. Careers in science and technology are not as valued as they should be (and once were) in the United States:

- Improving the social status of engineers and scientists will increase the number of people who pursue careers in these areas.
- The number of trained scientists and engineers may not be adequate if foreign-born scientists begin to return home or stay home instead of seeking education and employment in the United States.
- The stresses and demands of technical development jobs may lead to burnout and early retirement.

Economic Considerations

1. There will a dynamic interaction between technology developments and the structure of the economy, each driving the other at the same time as it is shaped by the other.
2. Opportunity (development of the Internet, for instance) and need (for, say, multidisciplinarity or life-long learning) will drive significant changes in education—for example, much improved online courses.
3. Environmental concerns should stimulate demand for technologies that increase efficiency, although this demand is more likely to come from industry than from consumers.
4. Opposition to nuclear power may abate as the economic and environmental risks of fossil fuels are understood.
5. There will be considerable demand for improved health technologies, and the direction of biomedical technology will influence the structure of health care itself:

- Health-care cost pressures are accelerating, in part as a result of technological advances, and those pressures are compounded by trends in demographics (aging of the population in the United States and other advanced economies).
- There will be demand for improved connectedness and communication in medical records and for the use of remote medical technologies.
- Reimbursement policies will impact the decisions to support new biomedical technologies.
- Advances in biomedical technology tend to increase costs, while advances in other technologies tend to decrease cost. Can this be changed?
- Technology can improve the health-care delivery system in areas such as record keeping, diagnostics, and so on, but this improvement may not decrease total cost if patients come to take for granted each new technical development.

- There will be growing demand for technologies that increase health care autonomy (self-monitoring, for example).
- Product liability may inhibit the incorporation of new technology into invasive biomedical products.

6. Although manufacturing may be declining relative to services, it is still a sector in which R&D and innovation are important to the national economy.

    - Economic forces will push manufacturing closer to the source of natural resources and to end-use markets. This may reverse the flow of manufacturing from the United States in certain industries.
    - The trend toward outsourcing by major corporations is being extended to R&D.

7. The U.S. economy promotes capitalization of innovation, which in turn facilitates the adoption of new technologies. This advantage of the United States will be increasingly shared by other economies as they adopt structures that can capitalize innovation.
8. Economic rewards would come from innovation in other areas as well:

    - Greater attention to standardized ways of presenting/mining information and retrieving information online.
    - Improvements to the modeling and simulation of economic systems.

International Context

1. The globalization of science and technology will have a major impact on the economy and the development of technology.

    - Multinational corporations already conduct borderless R&D.
    - A greater proportion of fundamental research may be conducted offshore.
    - Dramatically lower costs can be realized by relying on highly educated populations in India and China instead of the United States.

2. Advances in other areas would have significant impacts, positive and negative:

    - *Communications standards.* International standards will encourage development because they will generate confidence that the development will be used.

- *Environmental regulations.* International agreements may encourage some development and inhibit other development, depending on the needs generated and the activities prohibited.
- *Intellectual property policies.* Lack of intellectual property protection will inhibit development.

3. More consistent intellectual property protection would encourage investment in R&D:

   - Country-to-country differences in patent policies are a problem, and increased harmonization is needed.
   - First-to-file patent policy followed by most other nations would create a bias in favor of larger organizations and might inhibit innovation. (Not all members of the subgroup agreed with this forecast.)
   - Inconsistent interpretation of fair use of databases impedes scientific research.

4. The European Union is becoming a second large, homogeneous market. It has more aggressive environmental regulation and is driving international standards that could become trade barriers.
5. International cultural differences are important. The United States excels in creative industries (e.g., software, entertainment), Japan in incremental improvements (e.g., consumer electronics).
6. A number of technological developments, including the following, will be either driven or impeded by national security concerns:

   - Encryption technology (impeded),
   - Commercial use of high-precision Global Positioning System technology (impeded and driven),
   - Electromagnetic wave absorbing coatings (impeded and driven),
   - High-performance computers (impeded),
   - Satellite technology (impeded and driven),
   - Explosives detection (driven), and
   - The Internet (driven).

7. There will heightened concern about dual-use technologies that can also be used by terrorists in weapons of mass destruction: biotechnology and nuclear technology, for example.

Plenary Discussion of Pull Factors

There was an extended discussion of the public response to technological change and research risks and what explains that response. Public opinion in

France, for example, is opposed to genetically modified organisms but is not very concerned about nuclear power, which provides 75 percent of France's electricity. U.S. public opinion turned against nuclear power after the Three Mile Island incident, although no one was injured, but it is not very concerned about genetically modified organisms. The German public is much more accepting of new medical devices than the U.S. public. There was consensus that most people do not understand statistics, leading to a gap between perceived and objective risk. For example, airline safety is much more stringently regulated than automobile safety, and many more people die in car accidents each year. Some workshop participants argued that the public is more rational than most scientists and engineers give it credit for. In the case of automobile safety, people have much more choice than they do in the case of airplane safety. Although the idea that scientists and engineers are rational and nonscientists are not is obviously simplistic, there was consensus that public understanding of science and technology should be increased.

Another discussion ensued, this one on the difficulty of agreeing on the facts when negotiating international treaties, such as those on controlling global warming, biosafety, and patents, and on what the scientific bases for international standards are or should be—for example, should the Kyoto Protocol be based on the findings of the Intergovernmental Panel on Climate Change? The U.S. approach to setting product standards, working through the American National Standards Institute, is more private and voluntary than the European approach.

Inequities such as the digital divide here in the United States and the implications of technology transfer for developing countries were discussed in the light of new threats of terrorism. The consensus was that the pressure for technology transfer will increase and that modernizing trends in many countries will make them more capable of incorporating technology; however, governmental constraints dictated by foreign policy or national security considerations will also be an important factor.

E

Recent Reports on Future Trends in Science and Technology

Michael McGeary

CONTENTS

INTRODUCTION	100
SUMMARIES	101
Council on Competitiveness, 101	
National Intelligence Council, 104	
U.S. Commission on National Security/21st Century, 108	
RAND Science and Technology Policy Institute, 112	
Industrial Research Institute, 117	
Battelle, 119	
RAND, 120	
RAND, 123	
CONCLUSION	127

INTRODUCTION

The purpose of this paper is to summarize the recent views of external organizations on future technological and industrial trends. "Recent" was taken to mean reports issued during the past 2 years. Eight reports from six organizations were identified. The organizations are Battelle, the Council on Competitiveness, the Industrial Research Institute (IRI), the National Intelligence Council (NIC), RAND, and the U.S. Commission on National Security/21st Century (also known as the Hart-Rudman Commission).

The reports were selected because they are comprehensive in scope. That excluded a number of reports that focused on particular technologies or technology areas (e.g., information technology, nanotechnology). In several cases (Battelle, the Council on Competitiveness, IRI, and RAND), the organizations had issued earlier reports on the subject, which could also have been used, although the most recent reports tend to incorporate the results of the earlier work.

It is interesting that the most comprehensive forecasting exercises were conducted in the national security area. Those include the reports by NIC and the Hart-Rudman Commission and a report by RAND done as input to the NIC report. There is a long history of policy planning, including technology policy, in the national security arena, probably because the federal government has overall responsibility in that area, including maintenance of the U.S. technology base and surveillance of technological threats to the United States. On the domestic side, no agency has overall responsibility, and there is little incentive to undertake comprehensive planning or forecasting. The lead organization in this area, the Council on Competitiveness, was created as a response to an external threat—Japan's economic inroads into American and world markets in the 1980s.

As much as possible, the reports are allowed to speak for themselves. The summaries often directly quote from them or are close paraphrases of their main points. Comparative analysis is reserved for the conclusion of this paper.

SUMMARIES

Council on Competitiveness

U.S. Competitiveness 2001[1] analyzes U.S. economic performance from 1985 to 2000—what drove U.S. prosperity in the 1990s and where economic performance fell short—and identifies the challenges facing U.S. leadership to sustain the nation's competitive advantage. Porter and van Opstal conclude with recommendations for keeping the United States competitive in the future.

Drivers of U.S. Prosperity from 1985 to 2000

Porter and van Opstal show that the extended economic expansion was built on strengths in all key components of economic growth:

- Post-1995 growth in GDP per capita reached quarter-century highs.
- Two-thirds of GDP growth was due to increases in productivity growth and capital stock per worker that, in turn, were driven by investment in and deployment of new technologies.

---

[1]Michael E. Porter and Debra van Opstal, U.S. Competitiveness 2001: Strengths, Vulnerabilities and Long-Term Priorities, Council on Competitiveness, Washington, D.C., 2001. Available online at <http://www.compete.org/pdf/competitiveness2001.pdf>.

- High productivity growth, along with supportive monetary policy, permitted full employment without inflation.
- Entrepreneurial activity—much of it in technologically intensive fields such as information technology and health—created millions of new businesses and jobs (an estimated one-third of new jobs between 1990 and 1997).
- Government fiscal discipline freed up capital for private investment.
- The United States led the world in patenting, the best single measure of innovation.
- Although the trade deficit was large, there were trade surpluses in the innovation-intensive sectors of the U.S. economy—advanced services, high-technology services, and licensing of intellectual property.

Where Economic Performance Fell Short

The authors go on to argue that the economic boom of the 1990s masked persistent areas of weakness in the U.S. economy that could undermine longer-term prosperity:

- Forty percent of U.S. households did not prosper for most of the 1990s.
- U.S. income inequality was the highest in the industrialized world, indicating a growing skills and education gap and a failure to make the most of the nation's human resources.
- National investment in frontier research lagged, and enrollments and degrees in science and engineering, outside the life sciences, began a downward trajectory, which limited U.S. capacity for innovation.
- Personal savings rates hit lows not seen since the Great Depression, which drove the current account deficit to record levels (more than 4 percent of GDP) and increased U.S. dependence on foreign capital.

Challenges to U.S. Leadership

An increased commitment to innovation is needed just for the United States to stay even, because other countries are increasing their capacity for innovation. The elements of innovative capacity—talent, technology, and capital—that powered U.S. leadership in cutting-edge technologies are now globally available:

- More nations are acquiring high-end innovation capabilities with concerted investments in research and development and technical talent.
- Other nations are developing fast-follower capabilities to rapidly commercialize innovation that originated elsewhere.
- The supply of scientists, engineers, and technicians is growing substantially faster abroad than in the United States.

- America's first-mover advantage in information technology is diminishing because of aggressive investment in and deployment of information technology (IT) by the rest of the world.

Recommendations

To increase the nation's standard of living in the long run, policy makers will have to invest in innovation, increase workforce skills, and strengthen regional clusters of innovation. The authors recommend that the United States undertake the following:

Lead in Science and Technology

- Increase federal investment in frontier research.
- Strengthen support for fundamental disciplines that have been neglected.
- Expand the pool of U.S. scientists and engineers by upgrading K-12 math and science education, broadening the science and engineering pipeline to include more women and minorities and increasing incentives for institutions of higher learning to increase the numbers of graduates in scientific, engineering, and technical fields.
- Modernize the nation's research infrastructure.

Boost Overall Workforce Skills

A world-class workforce is needed to sustain global competitiveness. The fastest growing and best-paid jobs require some level of postsecondary education, but there is evidence that demand for education and skills is outstripping supply while a significant percentage of the population is not prepared. The nation must invest in education and training to maximize the potential of every citizen to sustain future economic growth.

- Improve math and science education.
- Provide access to information technology to all students.
- Raise postsecondary enrollment rates for underrepresented minorities.
- Increase access to higher education for students from low-income households.
- Extend training opportunities to more workers.

Address Changing Demographics

Nearly 30 percent of the population will be at or over retirement age by 2030, leaving behind a smaller and less experienced workforce.

- Bring more citizens into the workforce by employing the under- and unemployed and raising workforce participation among older workers.
- Increase productivity per worker by increasing investment in technology, training, and education.

Strengthen Regional Clusters of Innovation

The locus of innovation is increasingly regional, because geographic concentrations or clusters of firms, suppliers, and related institutions in particular fields constitute networks of technologies, resources, information, and talent that foster innovation.

- Extend the focus of competitiveness and innovation policy to the regional level.
- Support regional leadership initiatives and organizations that enhance and mobilize cluster assets.
- Identify best policy practices in cluster development.

National Intelligence Council

Global Trends 2015[2] attempts to summarize changes in the world over the next 15 years that national security policy makers should take into account, based on the expertise and analyses of a broad range of nongovernmental organizations and experts.

The effort began in late 1999, when several dozen government and nongovernmental experts in a wide range of fields participated in two workshops to identify the major factors that will drive global change over the next 15 years. The workshops were followed by a series of more than a dozen conferences at a number of government and private research institutions on specific topics and by discussions with outside experts. The results are synthesized in this 85-page report (the HTML version on the Web, without graphics, is 45 pages long).

The report identifies seven global drivers and related trends that will shape the international system in 2015: demographics; natural resources and the environment; science and technology; the global economy and globalization; national and international governance; future conflict; and the role of the United States. It is acknowledged that no single driver or trend will dominate, each driver will affect different regions and countries differently, and in some cases, the drivers may work at cross-purposes rather than be mutually reinforcing.

---

[2]National Intelligence Council, Global Trends 2015: A Dialogue About the Future with Nongovernment Experts, U.S. Government Printing Office, Washington, D.C., 2000, 85 pages. Available online at <http://www.cia.gov/cia/publications/globaltrends2015/index.html>.

Several drivers are given more prominence in *Global Trends 2015* than in an earlier effort, *Global Trends 2010*, published in 1997. For example, globalization is considered an even more powerful driver: "GT 2015 sees international economic dynamics—including developments in the World Trade Organization—and the spread of information technology as having much greater influence than portrayed in GT 2010" (p. 6). Also, "GT 2015 includes a more careful examination of the likely role of science and technology as a driver of global developments. In addition to the growing significance of information technology, biotechnology and other technologies carry much more weight in the present assessment" (p. 7). And, "GT 2015 emphasizes interactions among the drivers. For example, we discuss the relationship between S&T, military developments, and the potential for conflict" (p. 7).

Demographics

The world population will increase from 6.1 billion in 2000 to 7.2 billion in 2015, but nearly all the increase (95 percent) will be in developing countries, mostly in urban areas. There will be little or no growth in most of the advanced economies. One result implies a need for substantially increased per-worker productivity: "In advanced economies—and a growing number of emerging market countries—declining birthrates and aging will combine to increase health care and pension costs while reducing the relative size of the working population, straining the social contract, and leaving significant shortfalls in the size and capacity of the work force" (p. 8). This situation also implies the need for continued immigration in high-income countries to relieve labor shortages. Reductions in health care costs would also help relieve the situation.

Natural Resources and the Environment

The report sees adequate energy, despite a nearly 50 percent increase in demand over the next 15 years, because of increased energy efficiency and large supplies of oil and gas that will be more efficiently extracted. But by 2015, nearly half the population of the world will live in water-stressed countries, and disputes over water could lead to conflict, especially if there are other tensions (e.g., in the Middle East). There will be adequate supplies of food, driven by advances in technology, although there will be problems with distribution in some areas. Environmental problems will persist and grow, although the report sees declining pressure on the environment caused by economic growth, because of technological advances (e.g., increased use of fuel cells and hybrid engines; better pollution controls; increased use of solar and wind power; more efficient energy use; and more efficient exploration for and extraction of natural gas).

Science and Technology

According to the report, development and diffusion of information technology (IT) and biotechnology will be significant global drivers. Breakthroughs in materials technology and discoveries in nanotechnology will also be important. It identifies two major trends, (1) integration of existing disciplines to form new ones and (2) lateral development of technology: "The integration of information technology, biotechnology, materials sciences, and nanotechnology will generate a dramatic increase in innovation. The effects will be profound on business and commerce, public health, and safety," and, "Older established technologies will continue 'sidewise' development into new markets and applications, for example, developing innovative applications for 'old' computer chips" (p. 32).

Information Technology

Many new IT-enabled devices and services will be developed and rapidly diffused during the next 15 years. Wireless global Internet access is technologically feasible. Some countries and population groups, however, will not benefit much from the information revolution.

Biotechnology

The report says that the biotechnology revolution will be in full swing by 2015, "with major achievements in combating disease, increasing food production, reducing pollution, and enhancing the quality of life" (p. 33). Among the areas in which developments will be significant by 2015: genomic profiling; biomedical engineering; therapy and drug development; genetic modification; and DNA identification. But these developments will be restricted to the West and the wealthy in other countries because of expense, and there will be substantial moral and religious controversies over some biotechnologies.

Trends in science and technology pose major uncertainties. Advances in science and technology may lead to dramatic breakthroughs in agriculture and health and provide leapfrog applications, such as wireless communication systems in less developed countries without landlines, but how or even whether such advances will benefit every nation and group is impossible to forecast. And, increasing reliance on computer networks makes U.S. communications systems vulnerable targets for rogue states, terrorists, and criminal groups, and advances in biotechnology, nanotechnology, and materials sciences provide opportunities for adversaries to conduct biological warfare or bioterrorism.

The Global Economy

The report (released in December 2000) sees a sustained period of global economic growth through 2015, although there will be periodic financial crises.

The growth will be driven by five factors: political pressures for higher living standards; improved macroeconomic policies to reduce inflation; rising trade and investment; diffusion and incorporation of IT that increases economic efficiencies; and increasingly dynamic private sectors in many emerging market economies (along with deregulation and privatization in Europe and Japan). This optimistic outlook depends on avoiding "potential brakes to growth"—if, for example, the U.S. economy suffers a sustained downturn; Europe and Japan fail to manage their demographic challenges; China and/or India fail to sustain high growth; emerging market countries fail to reform their financial institutions; or global energy supplies are disrupted in a major way (p. 39).

National and International Governance

Globalization, including "greater and freer flow of information, capital, goods, services, people, and the diffusion of power to nonstate actors of all kinds," will threaten the capacity and authority of most governments and create a demand for greater international cooperation on transnational issues (p. 38). National governments will have increasingly less control of flows of information, technology, diseases, migrants, armaments, and financial transactions across their borders, and there will be increasing pressures for international cooperation. Transnational private-sector organizations, both private and nonprofit, will play a greater role.

Future Conflict

The risk of war among developed nations will be low (but more destructive if it does occur). The greatest potential for conflict will stem from internal disputes: "Experts agree that the United States, with its decisive edge in both information and weapons technology, will remain the dominant military power during the next 15 years. Further bolstering the strong position of the United States are its unparalleled economic power, its university system, and its investment in research and development—half of the total spent annually by the advanced industrial world" (p. 56). Because of this overwhelming edge in military power, adversaries can be expected to pursue other ways to threaten the United States, such as terrorism and attacks on critical infrastructure (communications, transportation, financial transactions, energy networks): "At the same time, the trend away from state-supported political terrorism and toward more diverse, free-wheeling, transnational networks—enabled by information technology—will continue" (p. 50). It is also more likely that state and nonstate adversaries will develop or acquire more sophisticated weaponry, including weapons of mass destruction, and use them.

Role of the United States

The United States will still have unparalleled global economic, technological, military, and diplomatic influence in the international system. It will be the leading proponent and beneficiary of globalization and will remain in the vanguard of the technological revolution in IT, biotechnology, and other fields; however, it will encounter resistance from allies as well as adversaries to what is considered U.S. hegemony. Diplomacy will be more complicated, with more developing nations (e.g., China, India, Brazil), regional organizations, multinational corporations, and nonprofit organizations to deal with. The United States will encounter greater difficulty in using its economic strength to achieve foreign policy goals. Meanwhile, other governments will invest more and more in science and technology and education to prosper in the global economy. Governments, including the United States, will also benefit from greater communication and cooperation between national security and domestic policy agencies to deal with transnational threats, for example, infectious diseases, bioterrorism, and criminal activities.

U.S. Commission on National Security/21st Century

New World Coming was an early report of the bipartisan U.S. Commission on National Security/21st Century,[3] chaired by Gary Hart and Warren Rudman, and created in 1998 in response to the view that U.S. national security policies and processes needed to be reexamined in light of changed circumstances. Those changes in the national security environment included not just the new geopolitical situation following the end of the Cold War but also significant technological, economic, social, and intellectual changes:

> Prominent among such changes is the information revolution and the accelerating discontinuities in a range of scientific and technological areas. Another is the increased integration of global finance and commerce, commonly called "globalization." Yet another is the ascendance of democratic governance and free-market economics to unprecedented levels, and another still the increasing importance of both multinational and nongovernmental actors in global affairs. The routines of professional life, too, in business, university, and other domains in advanced countries have been affected by the combination of new technologies and new management techniques. The internal cultures of organizations have been changing, usually in ways that make them more efficient and effective.[4]

---

[3]U.S. Commission on National Security/21st Century, New World Coming: American Security in the 21st Century: Supporting Research and Analysis, Phase I Report on the Emerging Global Security Environment for the First Quarter of the 21st Century, Washington, D.C., September 15, 1999, 145 pages. Available online at <http://www.nssg.gov/Reports/reports.htm>.

[4]Charles G. Boyd, Executive Director of the Commission, in the preface to Road Map for National Security: Imperative for Change, Phase III report of the commission, March 15, 2001, p. ix.

In its final, Phase III report in April 2001, the Hart-Rudman Commission concluded, among other things, that the U.S. systems of basic research and education were in crisis while other countries were expanding their investments. The commission recommended doubling the federal research and development budget by 2010 and increasing the competition for those funds on the basis of merit. It also recommended elevating the role of the President's science advisor to oversee R&D expansion and other critical tasks, such as reviving the national laboratory systems and instituting better stewardship of the science and technology assets of the nation. And it recommended passage of a new national security science and technology education act to produce more scientists and engineers and qualified teachers in science and math.[5]

The commission began its work by trying to understand how the world would evolve over the next 25 years. The Phase I report included a volume of supporting research and analysis prepared by a national security study group, a set of national security scholars and practitioners formed to provide the commission with research and analytical support.

The first chapter analyzes global dynamics in science and technology, economics, governance, and national defense from 2000 to 2025.

Science and Technology

The Phase I report begins by acknowledging that scientific advances and technological innovations in the 21st century will cause social and political discontinuities that cannot be foreseen. Nevertheless, the report goes on to look at trends that can be observed and their implications.

In asking what technologies will emerge over the next 25 years, the report begins by noting that a major characteristic will be the increasingly smaller scale of new technology. For most of the 20th century, increasing scale led to higher efficiency and performance. "Now, however, miniaturization, adaptability, and speed are primary traits. Ever more capacity is being placed on tiny silicon wafers, and we are beginning to mimic the molecular assembly capabilities of biological systems" (p. 6). The report goes on to identify three technology areas in which the most important innovations of the next 25 years will occur, including combinations of the three: information technology, biotechnology, and microelectromechanics (MEM).

According to the report, there are also indications that nanotechnology will develop rapidly and have potentially revolutionary impacts, especially at the intersection of information technologies and biotechnologies. In this area and oth-

---

[5]The commission also made a number of other recommendations to ensure national security, redesign key institutions of the executive branch, overhaul the federal military and civilian personnel systems, and reorganize Congress's role in national security matters.

ers, however, such advances are not likely to have much impact in the next 25 years. After a major innovation, it takes most of a 25-year period to develop products and create the production, transportation, and marketing infrastructures needed to make a difference.

The report spends a number of pages on the challenges posed by new technologies, as well as their benefits. One challenge is to adjust to the ever faster pace of change caused by technological innovations. New technologies will also raise difficult and contentious legal, moral, and religious issues. They may even affect national identity and eventually change the nature of government itself.

Global Economics

In this section, the report looks at structural changes in the new global economy. These include an explosion in the volume of international capital flows to developing countries in the 1990s, especially from private sources, and dramatic changes in production, both enabled by advances in information technology. This has led to a new kind of economic integration. First, the ratio of trade to world GDP is historically high, reflecting greater interaction among national economies. Second, trade is shifting from manufacturing to services and involving many more countries. Third, multinational corporations are creating truly global production networks. Fourth, stock markets have sprung up around the world and provide a way to accumulate savings and make investments based on market criteria. Fifth, international and multilateral institutions are playing important and expanding roles. Sixth, there are rising expectations around the world pressuring governments to reduce impediments to global economic integration.

The report then reviews the reasons for resistance to global economic integration that slow or even prevent change (even in the United States) and lead to uneven progress in globalization. Other present or future events disrupting globalization over the next 25 years could include civil war and international conflict, war, a major disruption in global energy markets, the AIDS epidemic in Africa and Asia, another major unexpected pandemic, or a world recession. Avoiding the last—world recession—depends on the continued strong performance of the U.S. economy.

The report concludes that, absent a major economic or political global crisis, the major trends in global finance, manufacturing, transportation, telecommunications, and trade described in the report will continue: "The cross-border web of global networks will deepen and widen as strategic alliances and affiliates increase their share of production and profits. The internationalization of production networks will also continue. But the speed at which other parts of the globe join the integrative process, and the inclusiveness with which countries are transformed, is likely to be uneven and in many cases much slower than anticipated"

(p. 29). Nevertheless, the global economic system could be multipolar by 2025, as China, India, and Brazil become substantial export markets.

The implications for U.S. national security include greater economic disparities among and within countries; increased interdependence among open economies, including the United States; exploitation of trade and private capital markets for parochial purposes by certain nations; and challenges to the identity of nations and therefore to the legitimacy of their governments.

U.S. Domestic Future

Part III of the report examines the social, technological, economic, and political trends shaping the future of the United States.

Social Trends

Unlike most other developed nations, the population of the United States is expected to grow over the next 25 years, from 273 million to 335 million. Despite the growth, the population will age substantially. Nearly 18 percent will be over 65 in 2025, and the ratio of those in the workforce to those in retirement will go from 3.9 to 1 in 1995 to 2.3 to 1 in 2025. Health-care costs will continue to rise. The racial and ethnic composition of America will change, as minorities increase from 28 percent to 38 percent of the population.

The United States has top universities, which enroll large numbers of foreign students. Non-U.S. citizens constitute a substantial share of the graduate students in many fields—the physical sciences, engineering, mathematics, and computer science—and many of them stay on to work in the United States. At the same time, K-12 students in the United States do not compare well internationally in science and mathematics (which is probably related to declining numbers of U.S. students who major in the physical sciences, mathematics, and engineering in college). A relatively high proportion of adults is illiterate.

Technology Trends

American preeminence in science and technology is expected to continue over the next 25 years, although global trends in technology will have their effects:

> American society is likely to remain in the forefront of the information revolution. Most of the seminal scientific research and technological innovation is done in the United States, and American society and the economy are very receptive to new innovations. Nevertheless, America's relative lead in this field will likely decrease as other societies adapt to the information age (p. 120).

Biotechnology is keeping America on the innovative side of the agricultural, medical, and chemical industries, which will maintain the United States as a dominant actor in these sectors for at least the next quarter century. However, it will also raise basic and divisive ethical questions such as those involving access to new and expensive technologies (p. 120).

Similarly, those countries that are able to fabricate and apply MEMS devices and nanotechnology are likely to have a significant economic and military edge over those who cannot (p. 121).

Economic Trends

The most dramatic effect of science and technology on the United States is likely to be their contributions to strong economic growth during much or all of the next 25 years. The size of the impact depends on the rate of growth and whether the gains in productivity from the information revolution will raise it to 3 percent or more.

The U.S. economy will become more internationalized. U.S. foreign trade as a share of GNP went from 11 percent to 24 percent during the last 20 years, and that trend should continue.

RAND Science and Technology Policy Institute

The recommendations set forth here are contained in *New Foundations for Growth*,[6] a report done for the National Science and Technology Council in the White House.[7] The report is based on a yearlong effort to identify ways to improve (or reduce barriers to) innovation in the United States. The recommendations in the report are based on a literature review, on a synthesis of ideas submitted in response to a call for papers that went out to hundreds of businesses, business organizations, and laboratories asking for ideas for improving innovation, and on inputs at several workshops.

The report was done at the RAND Science and Technology Policy Institute, a National Science Foundation (NSF)-funded research and development center that supports the Office of Science and Technology Policy (OSTP) and other federal agencies with objective outside analysis of science and technology policy issues.

The approach of the report is to look at steps that affect the U.S. "national innovation system," a concept named by Richard R. Nelson in 1993.[8] The Ameri-

---

[6]Stephen W. Popper and Caroline S. Wagner, New Foundations for Growth: The U.S. Innovation System Today and Tomorrow, An Executive Summary, MR-1338.0/1-OSTP, Science and Technology Policy Institute, RAND, Washington, D.C., January 2001. Available online at <http://www.rand.org/publications/MR/MR1338.0/MR1338.0.pdf>.

[7]The review and sign-off of the full report by the Office of Science and Technology Policy (OSTP) was held up by the change in administrations.

[8]Richard R. Nelson, ed., National Innovation Systems: A Comparative Analysis, Oxford University Press, New York, 1993.

can innovation system consists of a complex network of interacting institutions—industry, government agencies and laboratories, research universities, and non-profit research institutions.

Private industry in the United States is currently very innovative and well able to capitalize on innovation in the marketplace. But industry does not constitute all aspects of the innovation system, especially those activities on which firms are not likely to recoup their investment. The public sector plays an important role, including direct and indirect assistance to the process of innovation and support of the infrastructure that enables economic innovation. Popper and Wagner list a number of types of direct assistance, including funding of research and development; protecting intellectual property; setting technical standards; agricultural and manufacturing extension services; and procurement actions. Indirect assistance includes regulation of the financial infrastructure; favorable fiscal policies; improving the education system; providing the national transportation and information infrastructures; and assisting trade.

In the face of strong global competition, government is playing a larger role in sponsoring partnerships among the institutions in the innovation system where interactions are weak. Partnering of various types appears to be one of the emerging characteristics of innovative activity and needs to be better understood. Government can provide funding in areas difficult for or of little interest to industry. In some sectors, government is a large buyer, which influences what products are developed. Finally, government sets and enforces the rules of the game and also regulates negative spillovers such as pollution and unsafe products.

Popper and Wagner present a series of recommendations related to the functioning of national innovation as a system: ensure inputs; improve the environment; facilitate communications; and understand better the dynamics driving the system.

Ensuring Adequate Inputs

Education and Training

- Improve the quality of K-12 education in general and raise the level of math and science education in particular.
- Expand options for access to science and technology education among groups currently underrepresented in the workforce of those fields.
- Increase opportunities for retraining in science and technology for the current workforce.
- Take measures to determine that resources and incentives are in place to ensure the output of a sufficient supply of technically trained professionals from institutions of higher education.

The Portfolio of Public Research

- Ensure adequate levels of public funding for fundamental science and engineering research.
- Make funding decisions a more informed process for assessing priorities and providing balance across fields in a manner commensurate with the complexity of the national innovation system.

Research and Experimentation Tax Credit

- Consider whether making the resarch and experimentation (R&E) tax credit permanent would be beneficial to the national innovation system and the larger economy.

Targeted Policies to Enhance Resources

- Evaluate the development of mechanisms to encourage investment in emerging technology sectors that currently receive limited venture capital funding and how such sectors and points of advantageous entry might be determined.

Maintaining a Favorable Environment

Intellectual Property Protection

- Consider what measures may be required to ensure that patent review processes maintain currency with new technology developments.
- Assess the effects of recent policy changes (such as the Bayh-Dole and Stevenson-Wydler Acts) on the flows and balance of government-funded research and their effects on private sector activities.

Standards

- Begin a systematic review of the process for setting technical standards considering both the potential importance and limitations of government involvement.
- Consider the role and process of standard setting as an aspect of U.S. trade policy.

Infrastructure

- Assess national needs for new measurement and testing systems that would create a benefit across industries.

- Examine federal investment priorities to ensure that public investments in infratechnologies are sufficient to sustain the growth and development of the national innovation system in desired directions.

Partnerships

- Evaluate the importance of various kinds of partnerships, as well as public-private consortia, in pursuit of innovative activity, determine when the public good would best be served by their coming into being, and consider how they may be fostered.
- Define clearly the boundaries for legal cooperation and research among firms in the private sector as well as between firms and the government.
- Consider what policy guidelines would be needed for informing the construction and operation of partnerships with a public component.

Improving Communications

Coordination Within the Public Sector

- Raise the awareness of federal agencies about issues affecting the national innovation system and their own roles within that system.
- Seek to define and identify best practice across federal agencies and promote learning and transfer of such practices to other settings.
- Seek opportunities to create or use existing forums and venues to foster discussion among federal agencies, between federal agencies and their state and local counterparts, and among government, industry, and academia on issues of common interest affecting the national innovation system.

Industry and Government

- Seek ways to recognize explicitly the de facto partnership and mutuality of interest between public and private sector institutions in support of the national innovation system and to enhance the complementarity of their activities.

Improving Understanding by the General Public

- Seek means to raise public awareness of the importance of innovative activity and what is required in the way of public actions to support that activity.
- Raise the prominence of formal awards for leadership in the field of technology development.

Maintaining Dynamism

Data for Better Understanding the National Innovation System

• Improve timely access to available government agency data on innovative activity and harmonize existing government databases.
• Increase incentives for agencies to collect data on innovation and technology use and transfer by means of special surveys and expansion of routine collections.
• Develop new measures and data categories to improve understanding of the innovation system and the interplay between public and private actions.

Anticipating Needs and Consequences

• Explore new means to assist in formulating policies that will be adaptive and robust to a variety of possible outcomes rather than static and restrictive.
• Explore new means to enhance foresight and forward thinking about developments in the national innovation system and the implications of its actions for the society and economy.

Measuring Performance in R&D

• Work to improve methods for measuring the long-term social and economic performance of investments in basic research.

International Dimensions

• Identify centers of excellence in science and technology to encourage linkages and leverage across national boundaries.
• Examine the global patenting system for ways to improve the efficiency of the patenting process.
• Identify ways that government can facilitate product and process standardization across national boundaries and determine when it might be appropriate to do so from the perspective of U.S. interests.

Popper and Wagner conclude by selecting five top recommendations:

1. Ensure adequate levels of public funding for all fields of fundamental science and engineering.
2. Consider the benefits of making the R&E tax credit permanent.
3. Examine the global patenting system.
4. Improve opportunities for training and retraining the science and technology workforce.

5. Raise awareness within the federal agencies of their role in providing the infrastructure for the national innovation system.

Industrial Research Institute

From 1996 to 1998, the Industrial Research Institute (IRI) undertook several studies of future changes in industrial R&D and the technological innovation process during the next 10 years. This article, "Industrial R&D in 2008,"[9] summarizes the outcome of an effort at the 1998 IRI annual meeting to understand how the industrial R&D laboratory would operate a decade later.

Larson reported that the other IRI studies had found that changes in industrial R&D would accelerate over the next decade. A planning exercise highlighted several drivers of change, the most important of which are information technology and globalization. It found that the industrial laboratory of 2008 would be different in several dimensions: people; technology intelligence; data acquisition/computational capabilities; and the innovation process.

People

R&D scientists and engineers of 2008 will become extremely customer-focused and self-motivated, with a deep understanding of the value of technology to the firm. They will also be more versatile, with multidisciplinary backgrounds and skills that are constantly being upgraded. They will be more risk-taking than their predecessors, always looking for new opportunities and applications for their work, by developing broad networks of contacts.

R&D leaders will be more fast-paced and adept at creating teams with the right skills and personalities for their mission. They will work to ensure quality and effectiveness by aligning technical teams with corporate or business-unit strategies. Company needs will be closely correlated with the needs and interests of research personnel through more effective reward systems. The role of the R&D leader will evolve to one of integrating technology throughout the enterprise.

Technical Intelligence

Technical intelligence will be fully integrated throughout the firm and far more comprehensive than today. Future use of technology intelligence will be more organized, with a core group of experts trained in collecting, analyzing, and applying needed information. The system will be based on close coordination

---

[9]Charles F. Larson, "Industrial R&D in 2008," Research•Technology•Management, Industrial Research Institute, Washington, D.C.: November-December 1998. Available online at <http://www.iriinc.org/web/Publications/cfl-industrial2008.cfm>.

Data Acquisition/Computational Capabilities

Technical work will be more efficient and effective, utilizing a wide variety of outside resources. R&D is becoming much more productive through extensive use of advanced computer hardware and software, which enable techniques for microminiature and combinatorial analyses in numerous fields and thus provide a comprehensive prediction of real-world performance. Increased outsourcing, partnerships, and alliances with expert companies and universities will become necessary, along with a significant increase in IT investment.

The Innovation Process

The need for a "smart" organization will force reeducation of the workforce, which will have to understand the importance of openness to new ideas. High-performance cross-functional teams will have multiple skills, aligned values and rewards, consistent work processes, lateral thinking, and smooth transition from concept to development to commercialization. Scientists will not work in "silos" but will be organized to encompass a common purpose that fosters customer-supplier-organization interactions. All R&D staff will spend a major portion of their time with customers, in their markets. Virtual global laboratories, already common in some companies, will make around-the-clock R&D a reality for most companies.

Laboratory of 2008: Summary

How does industrial R&D of 2008 compare with that of the present? For the most part, the differences will be quite dramatic because the information age (which has already had a profound impact) promises to accelerate the rate of change in R&D organizations.

First, information technology will have a profound impact on the way R&D is conducted. Second, teams will be the norm, but the need for individual ideas and creativity will be more important than ever. Researchers will need to be even more adaptable to change and constantly enhancing their skills through knowledge-based programs. Third, the intellectual capacity of the organization must grow to compete in an intensely competitive global marketplace. Technical intelligence must be collected, protected, and applied by the entire organization. Finally, the innovation system must be seamless throughout the organization, driven by a vision of where the corporation wants to go, nurtured with a strategy of how it wants to get there, and managed by persons with a solid understanding of technology as well as business.

Battelle

According to *Top Strategic Technologies for 2020*,[10] the most important technological trends that will shape business and the world over the next two decades are the following:

- *Genetic-based medical and health care.* We will witness an explosion of medical technology originating from genetic research, giving us the ability to detect and correct many genetic-based diseases before they arise, possibly even in the womb. And a wide range of new pharmaceuticals that originated from genetic research will come onto the market, leading to treatments, cures, and preventive measures for a host of ailments.
- *High-power energy packages.* Developments such as highly advanced batteries, inexpensive fuel cells, and microgenerators of electricity will make many of our electronic projects and appliances highly mobile. These new, high-power, distributed energy systems will provide backup, if not primary, energy sources for appliances, homes, and vehicles. We'll also see further improvements in batteries—perhaps linked with solar power—and small generators fueled by natural gas.
- *GrinTech (green integrated technology).* Global crowding, fears of global climate change, and mountains of garbage will thrust environmental concerns to the forefront of consumer and industry attention around the world. The integration of advanced sensors, new materials, computer systems, energy systems, and manufacturing technologies will be used to eliminate waste and make projects completely recyclable. GrinTech will be especially important in agriculture, mining, manufacturing, and transportation systems.
- *Omnipresent computing.* We will be in constant contact with very miniature, wireless, highly mobile, powerful, and highly personalized computing with network access. Such computers may first appear on the market as watches or jewelry with the power of a computer and cellular phone. Later, we will have computers embedded in our clothing and possibly implanted under our skin.
- *Nanomachines.* Microscopic machines, measured in atoms rather than millimeters, will revolutionize several industries and will perform a wide range of jobs for us—from heating our homes to curing cancer. Nanomachines could be used to deliver drugs to highly localized places in the body, to clean arteries, and to repair the heart, brain, and other organs without surgery.

---

[10]Battelle, Top 10 Strategic Technologies for 2020, Battelle Annual Report, 1999, p. 27. Available online at <http://www.battelle.org/ar99/ar99.pdf>.

- *Personalized public transportation.* The continuing growth of cities will further stress our transportation infrastructure. Traffic jams and road rage will decline substantially as people drive their cars to remote parking areas and take highly advanced—and comfortable—trains into central cities and between cities. Yet an aging population with concerns about safety, convenience, and independence will demand personal vehicles. The challenge will be to integrate many individual cars within a coordinated and optimized public transportation network.
- *Designer foods and crops.* Through genetic engineering, researchers will develop crops that resist diseases and pests, greatly reducing the need for pesticides and other chemicals. Battelle predicts that most food sold in supermarkets will come from genetically engineered fruits, vegetables, and livestock. And nearly all cotton and wool for our clothing will be genetically engineered.
- *Intelligent goods and appliances.* Advances in quantum computing will lead to smaller, more powerful computers and electronics that will add amazing intelligence to appliances and other projects. These projects will likely include telephones with extensive phone directories, intelligent food packaging that tells your oven how to cook the food inside, refrigerators that help make out your shopping list and tell where to get the best price on the food you need, and maybe even a toaster that won't burn your toast.
- *Worldwide inexpensive and safe water.* Within the next 20 years, clean drinking water could become an expensive commodity. However, before water shortages become critical, technology will answer the challenge with advanced filtering, processing, and delivery of potable water. Desalination of water and water extraction from the air are two possibilities.
- *Super senses.* One of the hot technologies today is virtual reality. In 20 years, we will be marveling over "enhanced reality." Using sensors and electronic or genetic technology, we will be able to implant devices that allow us to hear better than ever before or see farther or in the dark.

RAND

The Global Technology Revolution[11] was based on work conducted by RAND's National Defense Research Institute for the National Intelligence Council's report *Global Trends 2015*.[12] The content consists of "a quick fore-

---

[11] Philip S. Anton, Richard Silberglitt, and James Schneider, The Global Technology Revolution: Bio/Nano/Materials Trends and Their Synergies with Information Technology by 2015, RAND, Santa Monica, Calif., 69 pages. Available online at <http://www.rand.org/publications/MR/MR1307/MR1307.pdf>.

[12] National Intelligence Council, Global Trends 2015: A Dialogue About the Future with Nongovernment Experts, U.S. Government Printing Office, Washington, D.C., 2000, 85 pages. Available online at <http://www.cia.gov/cia/publications/globaltrends2015/index.html>.

APPENDIX E   *121*

sight into global technology trends in biotechnology, nanotechnology, and materials technology and their implications for information technology and the world in 2015" (p. iii). The foresight exercise considers potential scientific and technical advances, enabled applications, potential barriers, and global implications. It also identifies wild-card technologies that are not as promising or less likely to mature by 2015, but which would have a significant impact if they were developed.

Technology Trends (Chapter Two)

Trends in the following technologies are discussed, with citations to a lengthy bibliography at the end of the report:

Genomics
    Genetic profiling and DNA analysis
    Cloning
    Genetically modified organisms
    Broader issues and implications of advances in genomics
Therapies and Drug Development
    Technology
    Broader issues and implications
Biomedical Engineering
    Organic tissues and organs
    Artificial materials, organs, and bionics
    Biomimetics and applied biology
    Surgical and diagnostic biotechnology
    Broader issues and implications
The Process of Materials Engineering
    Concept/materials design
    Materials selection, preparation, and fabrication
    Processing, properties, and performance
    Product/application
Smart Materials
    Technology
    Broader issues and implications
Self-Assembly
    Technology
    Broader issues and implications
Rapid Prototyping
    Technology
    Broader issues and implications
Buildings
Transportation

Energy systems
New materials
Nanomaterials
Nanotechnology
    Nanofabricated computation devices
    Biomolecular devices and molecular electronics
    Broader issues and implications
Integrated Microsystems and MEMS
    Smart systems-on-a-chip (and integration of optical and electronics components)
    Micro/nanoscale instrumentation and measurement technology
    Broader issues and implications
Molecular Manufacturing and Nanorobots
    Technology
    Broader issues and implications

Of these trends, the authors expect certain technologies to have the most promise for significant global effects, although uncertainty is high. In biotechnology, better disease control, custom drugs, gene therapy, age mitigation and reversal, memory drugs, prosthetics, bionic implants, animal transplants, and many other advances should continue to increase the human life span and perhaps performance. And by 2015, it may be possible to use genetic engineering techniques to "improve" the human species and clone humans. The substantial controversy over the use of such techniques should be in full swing by 2015.

In the area of materials, devices, and manufacturing, the most promising advances will be in smart materials, agile manufacturing, nanofabricated semiconductors, and integrated microsystems.

The technology wild cards, advances that are unlikely to have much impact by 2015, include novel nanoscale computers using quantum effect or molecular electronics, molecular manufacturing, and self-assembly methods, including biological approaches.

Discussion (Chapter Three)

The authors say their descriptions of technology trends give some indication of what might happen based on current movements and progress, but they acknowledge that the progress in and effect of those trends will be affected by enablers and barriers. They present three graphics showing high-growth and low-growth paths for three technologies—genetically modified foods, smart materials, and nanotechnology—and discuss key enabling factors and key barriers in each case.

Meta Trends

They also present a number of meta trends based on reviewing the technology trends discussed above and discussions in the open literature:

- Increasingly multidisciplinary nature of technology, in which technology trends are enabled by the contributions of two or more intersecting technologies;
- Accelerating pace of technological advance and change;
- Accelerating social and ethical concerns;
- Increasing need for educational breadth and depth;
- Longer life spans;
- Increasing threats to privacy;
- Continued globalization; and
- Effects of international competition on technology development.

Technology Revolution?

After discussing the possibilities for additive or even synergistic effects from simultaneous progress of multiple technologies and applications and the highly interactive nature of trend effects (e.g., social, economic, political, public opinion, environmental), the authors propose the possibility that the world is experiencing a multidisciplinary technology revolution, going beyond the agricultural, industrial, and information revolutions of the past. They provide a table (p. 46) on which Table E.1 is based.

As part of the technology revolution, the overall workforce will probably have to contribute to and understand increasingly interdisciplinary activities. Consumers and citizens should have a basic understanding of technology to make informed decisions. Scientists, engineers, and technologists will have a greater responsibility to understand and communicate the benefits and risks of technological innovations. Technology workers will probably need a deeper interdisciplinary education to enable teaming and understand when to bring in specialists from other disciplines. They will need to keep their skills current. Truly multidisciplinary teams will be needed for progress in some R&D areas. The old paradigm of hierarchical relations of technology is being replaced with one where a team searches for solutions in multiple disciplines. Finally, the technology revolution is changing the way people interact and live and work.

RAND

Science and Technology Issues of National Importance[13] is a draft of a report on the outcome of a suite of projects designed to aid the incoming director

---

[13]RAND, Science and Technology Issues of National Importance, DRR-2486/5-S&TPI, Santa Monica, Calif., April 2001, 77 pages.

TABLE E.1 The Technology Revolution: Trend Paths, Meta Trends, and "Tickets"

	Past Technology	Present Technology	Future Technology
Trend paths	Metals and traditional ceramics Engineering and biology separate Selective breeding Small-scale integration Micron plus lithography Mainframe Stand-alone computers	Composites and polymers Biomaterials Genetic insertion Very-large-scale integration Submicron lithography Personal computer Internet-connected machines	Smart materials Bioengineering Genetic engineering Ultra/gigascale integration Nanoassembly Microappliances Appliance and assistant networks
Meta trends	Single disciplinary Macro systems Local Physical	Dual/hierarchically disciplinary Microsystems Regional Information	Multidisciplinary Nano systems Global Knowledge
"Tickets" to the technology revolution	Trade schools General college Locally resourced projects Capital ($)	Highly specialized training Specialized degree Locally resourced components Increased capital ($$)	Multidisciplinary training Multidisciplinary degrees Products tailored to local resources Mixed

and staff of the Bush administration's Office of Science and Technology Policy. After it has undergone RAND's peer review process, the final version will become a RAND Issue Paper.

First, a team from the RAND Science & Technology Policy Institute worked with the Washington Advisory Group, a bipartisan organization of former senior federal S&T officials, including several presidential science advisors, to identify more than 50 science and technology policy issues of possible importance to the new administration. Second, an external advisory panel consisting of Erich Bloch, Ed David, Steve Dorman, Arati Prabakhar, Frank Press, and Robert White, went over the list and prioritized the issues. The result is 10 science and technology issues on which the new administration should focus policy attention.

Safety and Security Issues

1. *Strengthening the national aviation system.* By 2015, 40 percent of the international commercial fleet will be built by non-U.S. companies, a doubling of market share since 1995. Meanwhile, the flight system is becoming overburdened. Third, the fatal accident rate has not declined for 20 years and the nonfatal accident rate and number of close calls are both growing.

2. *Reviewing U.S. export controls on sensitive technologies.* Currently, the controls are "a patchwork of outdated laws and regulations that appear to be onerous, ineffective, and poorly suited to modern conditions. New approaches that balance economic and security concerns might better serve U.S. interests."

3. *Reassessing national missile defense options.* "Proposals for a new national missile defense pose difficult scientific and technical issues quite apart from political and military considerations." The President's science adviser will probably be asked to examine the principal technical concerns.

Continuing Challenges to America

4. *Rethinking global climate change policy.* "Evidence is mounting that greenhouse gases are changing the earth's climate. Numerous alternative energy technologies show great promise for reducing the human impact on global climate without causing adverse economic impacts. However, the Kyoto Protocol's targets and timetables for reducing greenhouse gas emissions around the world are producing stalemate rather than progress. A new, more flexible and adaptive approach appears in order."

5. *Anticipating energy crises.* "The U.S. is facing energy shortages and price rises as well as questions of long-term strategy. Underlying these issues is a deeper set of infrastructure problems that span the energy spectrum. Accordingly, better monitoring and planning are needed to help the U.S. cope with current and likely future energy crises."

6. *Improving education research.* "The federal government is the predominant force in education research. As many states adopt reforms and infuse massive new resources into their public schools, the federal government has a unique opportunity to strengthen the scientific research base in education. A more solid research base could, in turn, help states and districts use their resources more productively."

New Challenges That Require Greater Government Attention

7. *Protecting critical infrastructures.* "Critical infrastructures are information networks, infrastructures, and other systems that are vital to economic well being, national security, and public safety. Although these are controlled primarily by the private sector, the 'public goods' nature of these private infrastructures suggests a role for government working with the private sector to protect such assets for society at large."

8. *Managing the capabilities of genomic technologies.* "Genomic technologies will confront the new administration with challenges no nation has yet faced. Human genetic research could soon offer capabilities never before possible.... Such capabilities will also raise serious questions. It will be important for the government to define the potential illegal use of genomic technologies and to provide adequate disincentives and safeguards against such use."

9. *Meeting the governance challenges posed by emerging technologies.* "More generally, the pace of technological development in many areas raises fundamental governance challenges. Some of the emerging technology-related challenges include safety protocols and trade rules for the commercial sale of genetically modified foods; privacy of information sent over wireless networks; taxation equity between Internet-based businesses and traditional businesses; and intellectual property protection not only for software but also for new 'business methods,' such as online shipping or marketing, and even for strings of genetic codes. Success in governing these emerging issues will depend on cooperation among state and local governments, international organizations, and private industry."

The Remaining Challenge

10. *Coordinating federal research priorities to best serve the public interest.* "Increasingly, science and technology investments are seen as central to U.S. national well being. Yet the large stake in federal investments—approximately $70 billion annually—is not managed as a coherent whole. To ensure balance across priorities, the administration should consider managing the federal research enterprise as an investment portfolio."

CONCLUSION

First, there are different types of reports. Some look at the supply side of technology and project current trends into the future. This type includes the Battelle list of the top 10 strategic technologies for 2020 and the RAND report on trends in biotechnology, nanotechnology, and materials technology and their synergy with information technology from 2000 to 2015. Others look at the demand side. For example, the RAND report on science and technology issues of national importance focuses on societal needs. RAND has also conducted demand-side studies, not summarized here, that focus on industrial needs in the shorter term (e.g., the next 10 years).[14]

The national security reports by NIC and the Hart-Rudman Commission are another type of demand study. They look at what the United States needs to ensure national security over the next 15 years, although the emphasis is on how technologies are creating new vulnerabilities in U.S. security such as bioterrorism or attacks on the communications infrastructure.

Still other studies, including those by the Council on Competitiveness and RAND's *Foundations for Growth*, focus on the continuing capacity of the U.S. national innovation system to be globally competitive rather than on the supply of (or the demand for) particular technologies. In this approach, trends in the health and performance of R&D institutions are the concern. This approach also highlights trends in the investment of other countries in scientific research, technological development, and education and training of scientists and engineers.

Second, there are some common themes. In those reports that talk about technologies, there is consensus that the main arenas of technological innovation in the United States will be in information technology and biotechnology. There is nearly as much agreement that advances in materials and in micro- and nanotechnology will also be globally important, although they may take longer than 15 to 25 years. More interestingly, the view is that the intersections of those technologies will be where the most innovative advances are made. There are several corollaries. One, innovation will more and more require true interdisciplinary work—not just assembling components from different technologies but designing entirely new products based on knowledge from several fields. Two, partnerships will be used more and more in science and technology innovation.

In those studies that focus on the capacity of the innovation system, there is consensus that the government plays a critical role. Although private R&D is increasing sharply, it does not provide all the elements needed for innovation. Some elements of the system—basic research, education and training, standards setting, and encouraging technology development in certain areas that are too

---

[14]Stephen W. Popper, Caroline S. Wagner, and Eric V. Larson, New Forces at Work: Industry Views Critical Technologies, RAND, RAND Europe, Coopers & Lybrand, and Technology Radar, 1998.

risky for individual companies to invest in—depend on government actions and resources.

There is also consensus that economic globalization will continue, which will require the United States to increase its capacity (e.g., invest in knowledge and people) to stay competitive.

F

Trends in the Economy and Industrial Strength

Kevin Finneran

CONTENTS

INTRODUCTION	129
INDUSTRY RESEARCH	130
Internal Capital, 131	
Venture Capital, 133	
Public Opinion, 134	
Federal R&D Spending, 135	
Regulation, 136	
INTELLECTUAL PROPERTY	138
HEALTH CARE FINANCE	139
EDUCATION	140
NEWS YOU CAN USE	141
USEFUL READING	141

INTRODUCTION

The view of technology's future is not very clear from the laboratory bench. Although progress in scientific and engineering research is the prime mover in the innovation trajectory, the pace and direction of innovation are also influenced by a number of nontechnological forces, including industry structure, capital markets, international politics, public opinion, and a variety of government policies on acquisition of products and services, research funding, regulation, intellectual property, education, trade, immigration, and a host of other areas. These forces rarely if ever dictate the path of innovation, but they play an important role in speeding or slowing progress and in shifting direction. One can see how the restructuring of the telecommunications industry facilitated the development of cellular telephones, how the abundance of venture capital in the 1990s accelerated progress in computer and information technology, how public anxiety about nuclear power slowed its development, how environmental rules influenced the path of development in automotive and energy technology, and how government support for graduate education has helped create the sophisticated specialists in research and product development that make the United States so successful.

Looking at these nontechnological forces alone will tell us little about the future of technology; besides, it's impossible to say exactly how these forces will evolve over time. But understanding how they are likely to interact with technological forces will give us a much more realistic picture of how technology will evolve. This paper does not aspire to present a comprehensive view of all the external forces that can shape technological development. Rather, it looks at some of the major social forces that are important to the general climate for innovation, and it explores a few individual cases to illustrate how a different mix of lesser factors can influence a specific technology or industry. One could quickly compile a list of other factors—say, trade policy or tax policy—that are important but are not even mentioned here. The landscape of innovation is too vast and various to allow technologies in any simple template. The goal of this paper is to stimulate participants to think broadly about social forces when they begin to focus in on the likely trajectory of a specific technology.

INDUSTRY RESEARCH

Because most innovation takes place in industry, it makes sense to begin there. Several important industries are in upheaval. Electric utilities are breaking up into distribution, transmission, generation, and energy management companies. Chemical companies are moving into agricultural biotechnology. Many small medical biotechnology firms have been formed in the past two decades, and it is not yet clear if they will remain independent or be acquired by the giant pharmaceutical companies when they begin to make commercial products. And surprises are possible. Celera Genomics, the private company that raced the federal government to map the human genome, now defines itself as an information company. The future of the communications market is also uncertain. There is no doubt that there will be a growing demand for broadband Internet connections, but it remains to be seen if one of the industries now providing these links—local phone companies, long-distance phone companies, cable TV providers, satellite TV companies, or perhaps a new entrant such as the electric utilities—will come to dominate. Health care has seen enormous growth in managed care in the past decade, but it's still not clear what the future will be. Although we are not likely to see soon a dramatically expanded government role in health care after the debacle of the Clinton health care plan, we should not rule this out over the long term. In the interim, government will play a critical role through its management of Medicare and Medicaid. In each of these cases, the outcome will have an effect on which technologies are developed and commercialized.

It should not need stating that the overall health of the U.S. economy will affect the pace of innovation across all industries and technologies. A strong economy increases the pool of capital available for the purchase of new technology and for investment by companies in R&D. The 1990s illustrate this point in microcosm. During the first half of the decade, the economy grew at an annual rate of 2.4 percent, and capital investment grew by 0.7 percent annually. When GDP grew by 4.3 percent a year in the latter half of the decade, capital investment soared to 1.3 percent of GDP. Much of this new capital spending went to information technology, stimulating growth and innovation in that sector. Many economists believe that this helped create what was then called the "new economy," in which IT was stimulating innovation and productivity growth in all industries.

Although most Americans were pleased with the performance of the economy in the late 1990s, not all of the news was good. In 1985, the United States spent more than 2.9 percent of GDP on R&D. That percentage fell to about 2.5 percent in 1994, and although it rose in the late 1990s, it was less than 2.8 percent in 1999. It is worth noting that the compound annual growth rate in R&D was 4.37 percent during the economic expansion of 1975 to 1980 and 4.39 percent during the expansion of 1982 to 1990, but it reached only 3.43 percent during the expansion of 1991 to 1999. In other words, the growth in R&D during the 1990s, though impressive, was not that high for a period of economic expansion. And if R&D spending was relatively low during a period of expansion, what can we expect during a contraction?

Internal Capital

The primary source of funding for product research is internal capital, and companies have been increasing their research budgets in recent years. During the late 1990s, while the federal R&D investment was declining, industry R&D was growing steadily. In the mid-1980s, industry and government spent about the same amount on R&D. By 1999, industry was spending twice as much as government. About two-thirds of industry spending is for product development and the remainder for basic and applied research. The percentage devoted to basic and applied research dipped slightly in the early 1990s, when growth was slow, but recovered quickly in the second half of the decade. Will that continue?

The most recent survey by the Industrial Research Institute conducted in the fall of 2000 found that companies plan to hold steady the percentage of sales allocated to R&D even though the economy has cooled somewhat. The survey also uncovered some other significant trends. Companies are increasing their R&D related to new business projects and reducing spending geared to existing businesses. They are increasing their participation in alliances and joint R&D ventures and decreasing their spending on directed basic research and precompetitive consortia. They are also reducing contact with the federal laboratories. Companies are increasingly looking outside their own walls for new technology. They expect that there will be more acquisition of capability through mergers and acquisitions, more licensing of technology to others, and more outsourcing of R&D to other companies. In theory at least, these developments should lead to enhanced efficiency in the use of research resources. During the 1980s, one explanation offered for Japanese companies' relative success compared with U.S. companies was that U.S. companies were afflicted with the "not invented here" syndrome, meaning that they failed to take advantage of research done by others. They now seem to be cured.

In thinking about the environment for technological innovation, it makes sense to pay particular attention to the high-technology industries where much of the progress occurs. One reason high-technology companies have been able to increase their R&D budgets is that sales and profits were so strong in the 1990s. The cash was available to invest in R&D, and demand was vigorous for new products. That's not the case today. Business spending for information technology, which was rising at an annual rate of 31.4 percent in the first quarter of 2000, declined at an annual rate of 6.4 percent in the first quarter of 2001. Economists argue over how much overcapacity for computer hardware, software, and communications equipment exists in industry, but with estimates in the $100 billion range, few expect demand to rebound quickly. Fortunately, the momentum of technological innovation will continue to improve the quality of new products, and companies will eventually want to replace outdated technology. That happens quickly with information technology (IT), but no one can say how quickly in the current economic environment. For the immediate future, however, cash will be tight in the IT industry, and this will have an effect on R&D spending. In fact, the very characteristics of the IT industry that enabled it to spend so generously on R&D in good times could make it particularly difficult to maintain R&D levels when sales are slack.

The marginal production cost of IT products such as software is extremely low, which means that when sales rise, profit increases even faster. But these very profitable products are built on a costly base of research and marketing. And because these products are susceptible to rapid obsolescence, the base must be maintained constantly. When sales are weak, as they are now, profits disappear quickly. Yahoo had its revenue drop 42 percent in one quarter, while expenses remained virtually unchanged. After earning an $87 million operating profit in the fourth quarter of 2000, Yahoo saw a $33 million loss in the first quarter of 2001. The story was similar at Cisco Systems during its first full quarter in 2001. Revenue fell 30 percent from the previous quarter, and operating profit before charges fell 95 percent. If the market for IT products remains weak for a long time, it will put severe pressure on the R&D budget. The chances are good that well-established companies such as Yahoo and Cisco will be able to maintain their R&D foundation during hard times, but it will be difficult for smaller companies, which have been a critical source of innovation. If these companies start to fail, the bigger companies could decide to use their resources to acquire smaller companies on the cheap rather than to invest in their own R&D.

Venture Capital

Venture capital grew enormously in the late 1990s, from less than $8 billion in 1995 to roughly $100 billion in 2000. So-called "angel" capital (direct investments by wealthy individuals) may have grown even faster. Reliable data are not available because of the private nature of these investments, but a 1998 estimate by the National Commission on Entrepreneurship put that year's total at $20 billion, whereas the venture capital investment was $14 billion. The equity market also grew apace. The total value of IPOs grew from about $4 billion in 1990 to more than $60 billion in 1999. U.S. capital markets have slowed recently, as reflected in the sagging stock market, but their overall health is strong. Still, we should not expect to see the breakneck rate of growth that characterized the late 1990s.

We also have to look at where this money was invested. In the first quarter of 2001, even as everyone was talking about the decline in value of information technology stocks, 35 percent of venture capital investment went to Internet-specific companies, 19 percent to computer software and services, 15 percent to communications and media, and 12 percent to semiconductors and electronics, according to the National Venture Capital Association. Only 7 percent went to medical and health companies and 4.5 percent to biotechnology.

Given all the recent news about the sequencing of the human genome and the accompanying potential for significant medical progress, one might expect a vast inflow of investment for research. That has not been the case. The share of venture capital going to biotechnology has been declining steadily during the past 5 years, and no turnaround is in sight. In spite of the widely acknowledged potential of biotechnology, it cannot promise the enormous short-term profits that many infotech companies have achieved. A biotech drug typically requires 10 to 15 years to develop, costs up to half a billion dollars, and must navigate a rigorous federal approval process. That's not the music most venture capitalists want to hear. Although the large pharmaceutical companies have the resources to develop biotech products, access to resources for the small companies is uncertain. The market research firm Pharmaprojects reports that there are 373 biotech companies that are developing only one drug. It is not clear how many of these firms will have the funds necessary to continue their research long enough to know if it could lead to a successful new product. The creation of many new biotech firms raises hopes for a cornucopia of innovation, but these hopes will be realized only if we are able to put resources in the hands of those with the breakthrough ideas.

Public Opinion

Public opinion is reflected in government policy and in the decisions that consumers make. Its importance can be seen particularly clearly in the area of agricultural biotechnology. Just ask the Europeans. During most of the 1990s European regulators were approving commercial production of genetically modified (GM) foods and consumers were buying them. However, in the late 1990s fear of GM products spread quickly throughout Europe, and by 1999 the European Council of Ministers was forced to institute a de facto moratorium on approvals of new GM products. The most recent Eurobarometer survey of public opinion found that two-thirds of Europeans would not buy GM fruits even if they tasted better than other varieties.

A 1999 Gallup poll found that while Europeans were campaigning vigorously against GM foods, half of Americans reported that they had heard little or nothing about the subject, and the majority were favorably disposed toward food biotechnology. However, 16 percent were strongly opposed. Awareness of GM foods has certainly grown since then, but there are still no signs of mass opposition. Nevertheless, the small group of committed opponents could be very influential if something happened to undermine public faith in food safety. In Europe it seems that minority opposition became a mass movement when the outbreak of mad cow disease and the badly handled government response shattered public faith in government's management of food safety. It didn't matter that mad cow disease had nothing to do with biotechnology. General anxiety about food safety and lack of faith in government protections became expressed as opposition to GM foods. The 1999 Gallup poll found that three-fourths of Americans are at least fairly confident that the Food and Drug Administration can ensure the safety of the food supply, and only 5 percent have no confidence. Still, the European experience demonstrates how quickly public faith can be lost, and a similar scenario is not out of the question in the United States. If it occurs, it will dramatically affect the development of U.S. agricultural biotechnology. In the meantime, the absence of a European market for GM foods is certain to discourage U.S. development of GM products.

Thus far, European opposition to agricultural biotechnology has not spread to medical biotechnology, but European industry is aware that it could. Paul Drayson, chairman of BioIndustry Britain, an industry association, has warned company leaders that they need to launch an ambitious public education program about the benefits of biotechnology or risk seeing medical biotechnology run into the same wall that has hampered progress in agriculture. Americans also seem quite willing to accept biotech medicines. The most often cited reason for the appeal of biotech medicine is that the benefit to consumers is usually obvious, and consumers are willing to accept a little risk in return for a clear benefit. The trouble with GM food is that consumers do not perceive any obvious benefit to themselves. They are not willing to accept much risk in return for foods with a longer shelf life or that can be grown with one herbicide rather than another. Part of the animus against GM foods is resentment of the agrochemical companies that produce them and seem to be the primary beneficiaries of the new technology.

Federal R&D Spending

Industry R&D spending is certainly the most important contributor to product innovation, but it would be a mistake to ignore the role of federal investment. A study by Francis Narin and colleagues found that during the 1993 to 1994 period, 73 percent of industry patent applications cited federally funded research, and this percentage was even higher for chemicals and for drugs and medicines. Because the government investment is particularly important in basic research, the effects of changes in federal spending may not be felt for 10 years or more. The National Research Council's Board on Science, Technology, and Economic Policy (STEP) has been studying the federal research budget's evolution to gauge its effect on specific research fields. Its July 2001 report provides a timely update.

The 1990s was a period of shifting priorities for the federal government, and the new priorities seem likely to persist for the foreseeable future. The end of the Cold War and the desire to reduce the federal budget deficit led to a significant decline in military research conducted by the Departments of Defense (DOD) and Energy (DOE) as well declines in research funding in other agencies. Between 1993 and 1997, research spending declined by 27.5 percent at DOD, by 13.3 percent at the Department of Interior, 6.2 percent at the Department of Agriculture, and 5.2 percent at DOE. Research spending at the National Institutes of Health (NIH) increased by 11 percent during the period.

The cuts fell particularly hard on the physical sciences, engineering, and mathematics. For example, federal support for electrical engineering and physics research fell by almost one-third in real terms. A few fields were able to offset their losses from DOD and DOE by picking up funding from other agencies. For example, support for computer sciences and for metallurgy and materials engineering rose by about one-fourth. The increased support for the life sciences was not distributed evenly. Support for medical sciences rose much more quickly than did support for the biological sciences.

Beginning with the 1998 budget, federal research spending began to rise significantly. Total research spending for 1998 was 4.5 percent above the 1993 level, and in 1999 it exceeded the 1993 level by 11.7 percent. The increasing budget continued the trend toward increased funding for the life sciences. Between 1993 and 1999, the life sciences' share of federal research rose from 40 to 46 percent, and the share going to the physical sciences and engineering fell from 37 to 31 percent. In 1999, support for physics, geological sciences, and chemical, electrical, and mechanical engineering was down at least 20 percent from 1993 levels. Chemical and mechanical engineering and geological sciences were down even from 1997 levels. In addition, support for astronomy, chemistry, and atmospheric sciences remained flat or declined between 1997 and 1999. Support for materials engineering, which had grown between 1993 and 1997, fell back to roughly 1993 levels by 1999.

The fields where federal support increased throughout the 1993 to 1999 period include aeronautical, astronautical, civil, and other engineering; biological, medical, and computer sciences; and oceanography. Fields that experienced reductions between 1993 and 1997 but recovered by 1999 include environmental biology, agricultural sciences, mathematics, social sciences, and psychology. Unless Congress or the agency managers have a sudden change of heart, we can expect the shift of federal research funding toward the life sciences, and toward NIH in particular, to continue.

The STEP board warns that although federal spending is contributing a declining share of the nation's total R&D expenditures, federally supported research is still a critical contributor to the nation's innovative capacity. Federal spending makes up 27 percent of all U.S. R&D and 49 percent of basic research. The board concludes that "reductions in federal funding of a field of 20 percent or more have a substantial impact unless there are compensating increases in funding from nonfederal sources, which does not appear to be the case the last few years." What's more, federal funding is usually more stable and has a longer time horizon, which is conducive to breakthrough research.

The STEP board also notes that maintaining a good balance of research among fields is particularly important because of the growing importance of cross-disciplinary work in vital fields such as bioinformatics, nanotechnology, and climate change. Because it is impossible to predict where breakthroughs will occur, it makes sense to attend to the well-being of the full spectrum of research fields. Advances in hot fields such as biomedicine and computer science will depend on progress in fields such as physics, chemistry, and the engineering disciplines, which have had declining federal support. The board argues that achieving a better balance in the federal research portfolio is important to maintaining the pace of innovation that the nation wants and expects.

Regulation

Many people in industry see a far more important federal role for regulation than for R&D. Certainly regulation has a more direct effect on day-to-day decisions, and whereas industry can (in theory at least) compensate for the deficiencies in federal R&D investments, it is helpless in the face of a regulatory barrier.

Although regulation seldom stops development of a technology, it can often slow it. For example, the Lewin Group, a health policy consulting firm, found that the Food and Drug Administration is very slow in approving hybrid products such as laser-activated drugs that combine device and drug technology into one treatment. The problem is that the various components of the treatment must be

APPENDIX F

reviewed by different FDA divisions, such as the Center for Devices and Radiological Health, the Center for Biologics Evaluation and Research, and the Center for Drug Evaluation and Research. The Lewin Group urged the FDA to improve communication and coordination among the centers. If the use of nanotechnology in medical devices begins to deliver on its promise, this will be particularly important.

Federal regulation can also stimulate R&D in specific technologies. The Public Utilities Regulatory Policies Act of 1978 (PURPA) required electric utilities to interconnect with independent power producers using alternative sources of energy such as wind and solar power and to purchase power from them at very favorable rates. This was a shot in the arm for small power producers that catalyzed R&D in numerous technologies, particularly wind power. Wind machine efficiency improved significantly in the subsequent two decades, and U.S. wind energy capacity now exceeds 2,500 megawatts, the equivalent of two large conventional plants.

PURPA is not popular with the utility industry. The Edison Electric Institute (EEI), the industry trade group, wants to see PURPA repealed, because it requires utilities to purchase power they may not need at above-market prices. EEI argues that even after 20 years, PURPA has yet to stimulate significant power production from renewable sources (less than 1 percent of U.S. electricity is generated by nonhydropower renewable sources) and that it shelters renewable energy technologies from market competition. EEI maintains that PURPA forces consumers to pay more for electricity without providing the stimulus necessary to spur the development of truly cost-effective renewable energy technologies. EEI sees the larger restructuring and deregulation of the utility industry that is already under way as the key to accelerating innovation.

The Electric Power Research Institute (EPRI), the industry's research arm, has developed a roadmap of how technology might evolve if regulations are revised in ways that give utilities more flexibility. EPRI wants to enlist computer technology to develop a much more reliable, electronically controlled distribution grid that would make it possible to introduce enhanced consumer control of power use, superconducting transmission to increase capacity and efficiency, smaller and lower-cost distributed generation technology, and local power storage. Because this industry is highly regulated and is subject to extensive environmental restrictions, the course of its development will be strongly influenced by government action at the state and federal levels. The nation is in the midst of a transition to a restructured and less regulated electric utility industry, and it is impossible to say at this time what it will look like. All that can be said is that the structure and economic incentives that do result will have a critical influence on whether we will see innovation in areas such as transmission and distribution, micropower generators, and novel storage systems.

Another area where government policy can be critical is in the allocation of the spectrum for wireless communication. In its October 2000 report *The*

Economic Impact of Third-Generation Wireless Technology, the President's Council of Economic Advisors (CEA) argues that the United States should allocate some of the spectrum available to third-generation (3G) wireless technology that will provide high-speed, mobile access to the Internet and other communications networks. These devices will transmit data at up to 2 megabits per second, about as fast as a cable modem, and will adhere to international standards that will make it possible to use the device anywhere in the world.

Several European countries have already allocated and auctioned spectrum for 3G use, but in the United States the three bands of spectrum being considered for 3G use are already used by analog cellular phone providers, the Department of Defense, fixed wireless providers, satellite broadcasters, school systems, and private video teleconferences. Although some of the spectrum now allocated to digital wireless telephone service could be used by its owners for 3G, the CEA report considers this unlikely, because it would make this bandwidth more scarce and therefore more expensive for voice phone service and would require replacing billions of dollars in capital stock such as transmission equipment. Besides, squeezing too much activity into this bandwidth could exhaust its capacity.

The government's decision on allocating new bandwidth for 3G should not be a make-or-break decision for this technology. However, it will affect the pace at which new services become available, the cost of those services, and the range of services that are offered. And it will inevitably affect other technologies that could use the same bandwidth.

INTELLECTUAL PROPERTY

In the summer 2000 edition of *Issues in Science and Technology*, U.S. patent commissioner Q. Todd Dickinson described how the Patent Office is keeping up with new developments in technology and updating its practices to ensure that they facilitate innovation. He pointed out that all U.S. patents granted since 1976 are now available on the Internet and that the Patent Office had in the previous 2 years hired more than 500 new examiners in its Technology Center, which examines software, computers, and business method applications. He also praised new legislation that requires publication of most patent applications within 18 months after the U.S. filing or priority date, unless the applicant states that no application has been filed for a patent in another country.

Dickinson concluded by explaining that what is really needed is global harmonization of procedural and substantive requirements of patents. Too much time and effort are wasted meeting divergent requirements. The World Intellectual Property Organization (WIPO) adopted a Patent Law Treaty in June 2000 aimed at harmonizing patent procedures. It will come into force once it is ratified by ten WIPO states. Harmonizing substantive requirements will be more difficult. A key issue is that the United States has a first-to-invent system, whereas the rest of the world uses a first-to-file approach. Earlier attempts to resolve substantive differences have failed, but Dickinson wants the United States to try again. Any progress that can be made in harmonizing the world's patent systems will accelerate innovation everywhere.

HEALTH CARE FINANCE

Both government and private insurance companies will play a vital role here. Because almost all health care is paid for either by government or employee health insurance, these institutions essentially are the market. What they're not willing to pay for, no one is likely to develop. David Lawrence, CEO of Kaiser Permanente, explains that there are two key concerns. He says that there is enormous consumer interest in home diagnostic technology, but the problem now is that almost none of the information collected with home devices makes it into the patient's medical record. This information could be extremely valuable, but unless we develop a way to make it easily available to physicians when they need it, consumers will soon realize that the devices cannot deliver what they promise. He also sees enormous potential for quality-of-life technology such as a device that would make it easier for people with cerebral palsy to communicate or the revolutionary new wheelchair developed by Dean Kamen that can travel over uneven surfaces and even go up and down stairs. If these devices can be purchased with government or insurance dollars, the demand for innovation will be strong. Reimbursement policies will also be important for new biomaterials and for nanotechnology.

The newly created Health Technology Center in San Francisco commissioned a survey of physicians to elicit their views on the use of information technology in their practice. Although 96 percent of respondents agreed that Internet-enabled technologies will make the practice of medicine easier and improve quality of care no later than 2003, only 34 percent use Internet-enabled sources for information about prescription medications, and only 7 percent have adopted automated systems for prescribing medications. (The Institute of Medicine report *To Err Is Human: Building a Safer Health System* recommended adoption of such systems.)

The respondents said that the greatest barriers to use of Internet-enabled services are "a lack of uniform standards for health information and the inability of current health information applications to communicate among themselves." The vast majority believe that the federal Health Care Finance Administration and the private insurance companies must take the lead in removing these barriers by requiring physicians to use the Internet for claims processing. This would force the health-care system to develop uniform standards and protocols for communicating information.

The need for standards for communicating information was reinforced by a workshop convened by the National Science Foundation and the Food and Drug Administration, "Home Care Technologies for the 21st Century." The workshop found vast potential for the introduction of new home care medical technologies but identified the reluctance of health insurers to pay for home care and the absence of infrastructure standards as formidable barriers to progress.

The federal government took the first step in this direction with the passage of the Health Insurance Portability and Accountability Act of 1996, which directed the Department of Health and Human Services (HHS) to standardize the way that health information is recorded and to develop rules to protect individual privacy. HHS has proposed standardized reporting practices that will simplify the sharing of data and rules for privacy protection that will inevitably make medical record keeping more complex. Computers and printers used for medical records will have to be physically secured, all software password protected, all transmissions encrypted, and electronic audit trails enabled that will identify everyone who has accessed the data. Health-care providers worry that the new requirements will be very expensive to implement and will undermine the information-sharing advantages of computerized patient records. Yet the public is very clear that it takes medical privacy very seriously. Reconciling the goals of enhancing the communication of medical information and preserving its privacy will be a contentious challenge.

EDUCATION

The 2001 Council on Competitiveness report *U.S. Competitiveness 2001* (summarized in Michael McGeary's paper for the committee, Appendix E) places particular emphasis on human resources for R&D and for production. It worries that the United States could find itself without the brainpower to develop and produce the technology it can envision. For example, the report notes that the number of undergraduate degrees awarded for engineering, math and computer sciences, and the physical sciences was stagnant or declining from 1985 into the late 1990s. Only in the life sciences did the number of degrees grow during this period. The picture was similar for graduate programs. Enrollment grew at a healthy pace in the life sciences but grew only slightly in engineering, math and computer sciences, and the physical sciences. This decline is linked to the decline in federal support for research in these fields. The reduction in research funding has been accompanied by a drop in the support available to graduate students in those fields.

One reason graduate programs did not actually shrink is the large number of foreign-born students who come to the United States for graduate study. The percentage of doctoral degrees in science and engineering awarded to foreign-born students grew from 35 percent in 1987 to 41 percent in 1997. Many of these graduates remain in the United States, but a significant number return home to their native countries.

At the same time as student interest in science and engineering careers seems to be waning, demand by employers for scientists and engineers is growing rapidly. The Department of Labor predicts that the number of new jobs requiring science, engineering, and technical training will increase by 51 percent between 1998 and 2008. That's four times the projected average rate of job growth. The fastest rate of growth will be in computer, mathematical, and operations research, an area in which the number of undergraduate degrees has declined significantly.

The short-term response to the growing demand for technically trained workers has been to increase the number of temporary visas available to noncitizens with the needed technical skills, but this is not an optimum solution. U.S. citizens worry that their jobs are going to noncitizens and that this strategy is really an effort to keep salaries low. In addition, these workers are acquiring valuable skills on the job, but there is no guarantee that they will be using them in the United States. The Council warns that the United States needs to be educating more of its young people in science and engineering if it is to maintain its innovative capacity. If it does not, U.S. companies will have to hire even more noncitizens or get the work done abroad.

NEWS YOU CAN USE

One cannot draw a straight line from any of these nontechnological factors to an eventual technological development. The most powerful influences are the large national and international economic forces that are impossible to predict or to link directly to individual technologies. Still, we cannot ignore them in considering where technology is likely to move. Although economic growth in the United States has slowed, the overall condition of the economy is strong. With companies apparently willing to invest in R&D and venture capital funds available to back new ideas, the general prospect for innovation is sunny.

When one begins to focus on specific industries or technologies, in each case a different mix of factors comes into play. There can be no easy generalizations, because each case will be different. Venture capital is critical in one case, of secondary importance in others, and irrelevant in yet another. The same is true for all the other factors. The only operable generalization is that it's wise to cast a large net when considering forces that will influence technological development and then to evaluate them carefully to see which are most important in the specific instance.

USEFUL READING

Board on Science, Technology, and Economic Policy, National Research Council, *Trends in Federal Support of Research and Graduate Education*, National Academy Press, Washington, D.C., July 2001.
Council of Economic Advisors, *The Economic Impact of Third-Generation Wireless Technology*, Washington, D.C., October 2000.
Industrial Research Institute, "R&D Trends Forecast for 2001," Washington, D.C., October 2000.

Kohn, Linda T., Janet M. Corrigan, and Molla S. Donaldson, eds., *To Err Is Human: Building a Safer Health System*, Committee on Quality of Health Care in America, Institute of Medicine, National Academy Press, Washington, D.C., 2000.

Narin, Francis, Kimberly Hamilton, and Dominic Olivastro, "Increasing Linkage Between U.S. Technology and Public Science," in *AAAS Science and Technology Policy Yearbook 1998*, Albert H. Teich, Stephen D. Nelson, and Celia McEnaney, eds., Washington, D.C., 1998.

Porter, Michael E., and Deborah van Opstal, *U.S. Competitiveness 2001*, Council on Competitiveness, Washington, D.C., 2001.

Winters, Jack, *Report of the Workshop on Home Care Technologies for the 21st Century*, Catholic University of America, Washington, D.C., 1999.

G
Innovation's Quickening Pace: Summary and Extrapolation of Frontiers of Science/Frontiers of Engineering Papers and Presentations, 1997-2000

James Schultz

CONTENTS

INTRODUCTION	144
Constraints and Limitations, 144	
TREND 1: COMPUTATION CONTINUES TO ADVANCE	145
TREND 2: A QUICKENING MEDICAL-GENETICS REVOLUTION	146
Genomic Medicine, 146	
Genetic Cures?, 148	
TREND 3: THE NANOTECHNOLOGY POTENTIAL	149
Tools of the Trade, 149	
Tiny Building Blocks, 150	
TREND 4: NATURE AS ENGINEERING TEMPLATE	153
Microscale Materials, 155	
TREND 5: THE MATURATION OF AUTONOMOUS MACHINES	156
ADDITIONAL TOPICS OF NOTE	159
Climate Change, 159	
Fluorescence Sensing, 160	
Neotectonics, 161	
Extrasolar Planets, 162	
Femtochemistry, 163	
Biometric Identification, 164	

INTRODUCTION

Scientific endeavor continues to accelerate rapidly on many fronts. Such a conclusion can be reasonably drawn from consideration of a series of symposia underwritten by the National Academy of Sciences (NAS) and the National Academy of Engineering (NAE). The symposia yearly bring together 100 of the nation's outstanding young scientists and engineers, ages 30 to 45, from industry, academia, and government to discuss pioneering research in various scientific, technological, and engineering fields and in industry sectors. Meetings are held not only in the United States, but also in China, Germany, and Japan, underscoring the now-ordinary nature of international scientific collaboration.

Following a competitive nomination and selection process, participation in the Frontiers of Science (FOS) and Frontiers of Engineering (FOE) symposia is by invitation. Attendees are selected from a pool of researchers at or under age 45 who have made significant contributions to science; they include recipients of the Sloan, Packard, and MacArthur fellowships; winners of the Waterman award; Beckman Young Investigators; and NSF Presidential Faculty Fellows. The symposia make public the thinking and work of some of the country's best and brightest, those now rising to positions of leadership within their institutions and industries.

Symposia presenters generally credit interdisciplinary approaches to fields of study as a means of providing new insights and techniques while significantly advancing specific disciplines. A survey of the symposia presentations indicates this cross-fertilization appears likely to intensify. For example, the use and integration of computation within all fields of research is now routine. The application of next-generation computation—from embedded microsensors and actuators within semi-intelligent machines and structures to the reverse engineering of biological systems (known as biomimetics) to enhanced modeling of cosmological, geological, and climatological phenomena—appears likely to facilitate additional advances, accelerating both basic understanding of fundamental processes and the efficacy of targeting applications and commercial spinoffs.

Constraints and Limitations

Given the availability of published papers and the timeliness of the research, the purview of this report is of necessity constrained to a span of not more than 4 years (although none of the year 2000 FOS papers had become available as of this writing, and neither are some 1999 FOS studies available.) The discussion herein is thus limited to a discussion of FOS presentations that occurred in 1997, 1998, and 1999 and of FOE presentations that occurred from 1997 through 2000. Presenters often limit themselves to the minutiae of their respective fields—an inclination to be expected, but not conducive to an aggregate synthesis of progress across multiple disciplines.

Dated reports of innovation are inevitably eclipsed by other, newer advances that tend to play out rapidly in either the scientific or popular press. Thus, this report is also dated in the scope and freshness of announced results. Nevertheless, these presentations do offer a glimpse into certain trends that underscores the acceleration of scientific enterprise. In general, the pace of innovation appears to be quickening, with multiple advances in multiple fields leading to yet greater technological acceleration. Nothing is ever certain, but if the rate of applied innovation continues to increase, even the most optimistic forecasts may prove in retrospect hesitating and timid. Although rearranged thematically and edited for clarity, the sections that follow contain material from the original papers published by the cited authors.

TREND 1: COMPUTATION CONTINUES TO ADVANCE

As computational power continues a seemingly inexorable advance, interest in and exploitation of new microprocessor architectures and software techniques remain strong. Computing is increasingly Internet-centric, with computer nodes distributed among desktops rather than in climate-controlled repositories. As Gharachorloo notes, administrators are finding it easy and inexpensive to scale up computational power by installing multiple independent systems. Although management issues inevitably arise, making more systems available simultaneously in a loosely clustered environment allows for incremental capacity expansion. The accelerating growth of the World Wide Web should continue to encourage the development and deployment of high-performance computers with high availability and incremental scalability. Hardware improvements are, however, proving easier to implement than is the software to support those improvements.[1]

Outside the purview of the FOS/FOE symposia are advances in biological and optical computing, both of which must be said to be in the very earliest stages of development. Nevertheless, their potential, either as stand-alone systems or integrated in some as-yet-unanticipated fashion, could have a substantial impact on future computer design and deployment. Even though robust real-world architectures for either have yet to be perfected, their promise, in terms of sheer computational power, is orders of magnitude beyond current serial-processing applications, and they should be considered as future contributors to advanced computing initiatives.

More promising still is quantum computation, which employs individual atoms, molecules, or photons in exploitation of quantum interference to solve oth-

---

[1]Frontiers of Engineering/1999. "Evolution of Large Multiprocessor Servers," Kourosh Gharachorloo, pp. 11-19. Many of the FOE papers cited in this report can be found at <http://www.nae.edu/nae/NAEFOE.nsf/weblinks/NAEW-4NLSEK?OpenDocument>.

erwise intractable problems. In principle, computers could be built to take advantage of genuine quantum phenomena such as entanglement and interference that have no classical analogue and that offer otherwise impossible capabilities and speeds. Computers that thrive on entangled quantum information could thus run exponentially faster than classical computers, say Brassard et al.[2]

Brassard et al. explain that quantum parallelism arises because a quantum operation acting on a superposition of inputs produces a superposition of outputs. The unit of quantum information is the quantum bit, or qubit. Classical bits can take a value of either 0 or 1, but qubits can be in a linear superposition of the two classical states.

The quartet assert that implementation of quantum computers presents a profound experimental challenge. Quantum computer hardware must satisfy fundamental constraints. First, qubits must interact very weakly with the environment to preserve their superpositions. Second, the qubits must interact very strongly with one another to make logic gates and transfer information. Lastly, the states of the qubits must be able to be initialized and read out with high efficiency. Although few physical systems can satisfy these seemingly conflicting requirements, a notable exception is a collection of charged atoms (ions) held in an electromagnetic trap. Here, each atom stores a qubit of information in a pair of internal electronic levels. Each atom's levels are well protected from environmental influences, which is why such energy levels also are used for atomic clocks.

For the moment, however, no large-scale quantum computation has been achieved in the laboratory. Nevertheless, several teams around the globe are working at small-scale prototypes, and quantum computing may be possible within the decade.

TREND 2: A QUICKENING MEDICAL-GENETICS REVOLUTION

Genomic Medicine

New kinds of diagnostic and therapeutic treatments will likely be derived from an enhanced understanding of the human genome. As Fields et al. see it, the emerging field of functional genomics—the term refers to a gene's inner workings and interplay with other genes—seeks to contribute to the elucidation of some fundamental questions. The three ask: How does the exact sequence of human DNA differ between individuals? What are the differences that result in disease or predisposition to disease? What is the specific role of each protein

---

[2]Frontiers of Science/1997. Gilles Brassard, Isaac Chuang, Seth Lloyd, and Christopher Monroe, at <http://www.pnas.org/cgi/content/full/95/19/11032>.

synthesized by a bacterial pathogen, by a model organism (e.g., *Escherichia coli*, yeast, the fruit fly, and the nematode), or by a human? How do proteins collaborate to perform the tasks required for life? Because not all genes are active in a given cell at a given time, which genes are used under which circumstances? How does this differential gene expression result in different types of cells and tissues in a multicellular organism?[3]

The trio point out that DNA arrays are being used to characterize human genetic variation. Extensive stretches of DNA sequence can be screened at once, and more than 4,000 variations have been found across the human genome. These small differences provide markers that can be used in subsequent studies to identify the genes responsible for particular traits and to analyze the sequences of important genes to uncover associations between specific genetic variations. Either predisposition to common diseases or the efficacy and safety of therapies can thus be iterated. Sophisticated tests should eventually allow the collection and analysis of the cellular and genetic information that should significantly augment biologic understanding, changing the way drugs are developed and the way diseases are diagnosed, prevented, and treated.

A number of studies are beginning to pinpoint both genetic predisposition and genetic function. For example, Uhl et al. say that precise identification of behavior-influencing candidate genes in humans and animals should lead to a more comprehensive understanding of the molecular/neurobiological underpinnings of complex behavioral disorders, as in animal studies that examine vulnerability to drug abuse. Classical human genetic studies also indicate significant genetic contributions to drug abuse, while genetic influences on human drug-abuse behaviors can be found in strain comparison, selective breeding, quantitative trait locus, and transgenic mouse studies. Testing genetic markers found within the human brain is also now possible.[4]

Identification of poorly understood, if complex, disease mechanisms may enable physicians to short-circuit the ways in which illness takes hold. The role of prions—proteinaceous infectious agents, molecules that cause transmissible neurodegenerative diseases in mammals—is now being evaluated, report Westaway et al. One class of prion protein was discovered during studies of experimental scrapie disease in rodents, while other varieties are associated with heritable cytoplasmic traits in yeast and fungi. All prion proteins are host-encoded and come in at least two varieties. The benign "cellular" form is referred to as PrPC, a molecule that is most probably present in all mammals and expressed on the surface of neurons.[5]

---

[3]Frontiers of Science/1998. Stanley Fields, Yuji Kohara, and David J. Lockhart, at <http://www.pnas.org/cgi/content/full/96/16/8825>.

[4]Frontiers of Science/1996. George R. Uhl, Lisa H. Gold, and Neil Risch, at < http://www.pnas.org/cgi/content/full/94/7/2785>.

[5]Frontiers of Science/1997. David Westaway, Glenn Telling, and Suzette Priola, at <http://www.pnas.org/cgi/content/full/95/19/11030>.

The three scientists contend that, while the prion hypothesis is not universally accepted, prions nevertheless may be gatekeepers controlling disease susceptibility. At the other extreme, prion proteins serve as the prototype for a new class of infectious pathogen and establish protein misfolding as a novel mechanism of disease pathogenesis, prompting the suggestion that simple organisms use prionlike mechanisms to switch physiological states and thereby adapt to new environments.

Genetic Cures?

Correcting defects at the genetic level should offer the most potent means of ensuring health and well-being. As defined by authors Kay et al., gene therapy is the introduction of nucleic acids into cells for the purpose of altering the course of a medical condition or disease. In general, with some exceptions, the nucleic acids are DNA molecules encoding gene products or proteins. Thus, the gene essentially can be considered a new pharmaceutical agent for treating many types of diseases. But because cells and organisms have developed powerful mechanisms to avoid the accumulation of extraneous genetic material, routine gene therapy is quite difficult, involving insertion of the appropriate gene into a target non-germ-cell tissue, such that an appropriate amount of gene product (usually a protein) is produced to correct a given malady.[6]

The three researchers assert that there are two primary candidates for gene transfer: viral and nonviral vectors. According to them, some believe that viruses will be most successful because they have evolved for millions of years to become efficient vesicles for transferring genetic material into cells, whereas others believe that the side effects of viruses and previous exposures will render the host resistant to transduction (gene transfer into the cell) and therefore preclude their long-term use in gene therapy. There are a number of additional viral vectors based on Epstein-Barr virus, herpes, simian virus 40, papilloma, nonhuman lentiviruses, and hepatitis viruses that are currently being evaluated in the laboratory.

Kay et al. remark that once a vector is designed, two general approaches are used for gene transfer: ex vivo, wherein cells are removed, genetically modified, and transplanted back into the same recipient, and in vivo, which is accomplished by transfer of genetic materials directly into the patient. The latter is preferable in most situations, because the complexity of the former method makes it less amenable to wide-scale application.

---

[6]Frontiers of Science/1997. Mark A. Kay, Dexi Liu, and Peter M. Hoogerbrugge, at <http://www.pnas.org/cgi/content/full/94/24/12744>.

TREND 3: THE NANOTECHNOLOGY POTENTIAL

Tools of the Trade

Building structures atom by atom, first to the molecular scale and then to the macro scale, is the defining characteristic of the emerging field of nanotechnology. Wiesendanger contends that a key nanotechnology tool is the scanning tunneling microscope (STM). In STM and related scanning-probe methods, he writes, a probe tip of atomic sharpness is brought within close proximity to the object under investigation until some physical signal can be measured that might originate from electronic, electrical, magnetic, optical, thermal, or other kinds of interactions between tip and sample. Point probing by a sharp tip allows one to receive local information about the physical, chemical, or biological state of a sample, which facilitates the investigation of site-specific properties.[7]

As Wiesendanger explains, to achieve high spatial resolution the distance between the probe tip and the sample is chosen to be smaller than the characteristic wavelength of the particular type of interaction acting between tip and sample. In the case of STM, that distance would be the electron wavelength, whereas for a scanning optical microscope it would be the optical wavelength. STM and related scanning-probe methods are therefore exceptional types of microscopes because they work without lenses, in contrast to optical and electron microscopes, and thus achieve superresolution.

He writes that, for strongly distance-dependent interactions, the dominant tip-sample interaction region can be as small as a few angstroms, thereby allowing the imaging of individual atoms and molecules on surfaces. Increasing the interaction strength between probe tip and sample in a controllable manner has become important for the fabrication of well-defined nanometer-scale structures. It has even become possible to synthesize artificial structures by sliding individual atoms and molecules on surfaces by means of the probe tip.

According to Wiesendanger, the controlled manipulation of matter at the scale of individual atoms and molecules may lead to new generations of nanoelectronic and mass storage devices. Scanning-probe instruments have also become powerful metrological devices that allow measurement of the distances between objects with extremely high accuracy. The precision of distance measurements offered by scanning-probe methods is also of great importance for a variety of sensor applications.

Scientists using scanning-probe microscopes and advanced optical methods are able to closely study single molecules. As Bai et al. note in their writings, the findings of these studies not only confirm the results expected from studies of bulk matter, but also provide substantially new information on the complexity of

---

[7]Frontiers of Science/1997. Roland Wiesendanger, at <http://www.pnas.org/cgi/content/full/94/24/12749>.

biomolecules or molecules in a structured environment.[8] The technique lays the groundwork for achieving the control of an individual molecule's motion. Ultimately, this work may lead to such practical applications as miniaturized sensors.

The four assert that studying single molecules is important because molecular individuality plays an important role even when molecular structure is complex. An intricate internal structure such as that found in a biomolecule, for example, results in a complex energy landscape. Alternatively, the molecule may be influenced by environmental factors that substantially change its behavior. Thus, the ability to distinguish different molecules under differing conditions and structures becomes crucial for understanding the system as a whole.

Biomolecules in living cells are one such example, the quartet explain. Even simple inorganic molecules on structured surfaces or in disordered systems, such as viscous liquids or glasses, provide situations in which molecular individuality matters. In all of these cases, the ability to study an individual molecule over time can give new insights unavailable by straightforward experiments on macroscopic populations of molecules. The new questions that single-molecule experiments pose move chemistry and physics into a realm more familiar to astronomers, who have direct observational knowledge of a single complex object such as the universe, and who must infer underlying rules and patterns. A single molecule under active control may well resemble the elegant engineered machinery rather than the "wild" molecules created by and found commonly in the natural world.

Tiny Building Blocks

As Wooley et al. point out, over the past decade, polymer chemistry has attained the sophistication necessary to produce macromolecules with accurate control of structure, composition, and properties over several length scales, from typical small-molecule, angstrom-scale resolution to nanometer dimensions and beyond.[9] Most recently, methods that allow for the preparation of polymeric materials with elaborate structures and functions—"bioinspired" materials—have been modeled from biological systems.

The four explain that, in biological systems, complexity is constructed by the ordering of polymer components (that is, polymers of amino acids, saccharides, and nucleic acids) through a combination of covalent bonds and weak interactions (such as hydrophobic, hydrogen bonding, and electrostatic interactions), with the process being controlled by the specific sequence compositions. An

---

[8]Frontiers of Science/1998. Chunli Bai, Chen Wang, X. Sunney Xie, and Peter G. Wolynes, at <http://www.pnas.org/cgi/content/full/96/20/11075>.

[9]Frontiers of Science/1999. Karen L. Wooley, Jeffrey S. Moore, Chi Wu, and Yulian Yang, at <http://www.pnas.org/cgi/content/full/97/21/11147>.

extension of these assembly concepts to synthetic macromolecules is primarily driven by the desire to create materials that resemble the structure and morphology of biological systems but that nevertheless possess the versatile compositions and properties of synthetic materials.

According to Zhou, nanotechnology's most effective building blocks may be extremely strong carbon nanotubes. He writes that few, if any, materials demonstrate as perfect a structure at the molecular level as does a single carbon nanotube, which exhibits excellent mechanical and thermal properties. A nanotube's aspect ratio and small diameter permit excellent imaging applications, and theoretical calculations and measurements performed on individual carbon nanotubes have demonstrated that their elastic modulus is as high as that of diamond. Indeed, Zhou asserts, if engineers could make a defect-free, carbon nanotube cable, such a cable that connected Earth and the Moon would be within the realm of possibility. Carbon materials could also be used in spacecraft, both for vehicle structures and for providing power from fuel cells or lithium-ion batteries based on nanomaterials.[10]

Architectures built from pure carbon units can result in new symmetries and structures with unique physical properties, according to Ajayan et al. The three write that a carbon nanotube can be considered the ultimate fiber—highly organized, near-ideal bonded-carbon structure. The organization of the hexagonal honeycomb carbon lattice into cylinders with helical arrangement of hexagonal arrays creates an unusual macromolecular structure that is by far the best carbon fiber ever made.[11]

Because of their intrinsic structural robustness and high electrical conductivity, nanotubes show great promise for applications, the trio maintain. Studies of field emissions have demonstrated transmission of large currents at low operating voltages. Nanotubes emit radiation coherently, indicating possible use in electron holography. Electron emissions from arrays of tubes have also been used to construct an actual device: a cathode ray tube lighting element. Now nanotube arrays can also be grown on glass substrates for field-emitting flat panel displays.

According to Ajayan et al., techniques for the robust attachment of multiwall nanotubes to the ends of atomic force microscope (AFM) cantilevers have been developed for use as scanning probe tips. Such tips have a number of advantages, including crashproof operation and the ability to image deep structures inaccessible to conventional tips. Such tips have been used as high-resolution probes in electrostatic force microscopy and in nanolithography for writing 10-nanometer-wide lines at relatively high speeds. Chemistry can be restricted to the ends of the

---

[10]Frontiers of Engineering/2000. "Science and Technology of Nanotube-Based Materials," Otto Z. Zhou, pp. 89-92.

[11]Frontiers of Science/1999. P.M. Ajayan, J.C. Charlier, and A.G. Rinzler, at <http://www.pnas.org/cgi/content/full/96/25/14199>.

nanotubes, where topological defects are present, and the nanotube tip can be functionalized with distinct chemical groups and used to demonstrate nanoscale imaging with the ability to discriminate the local chemistry of the surface being probed.

In addition, note the three researchers, nanotubes can be considered as the ultimate carbon fiber and may one day replace existing micron-size carbon fibers in composites. Freestanding films of purified nanotubes have also shown promise for application as the anode intercalation host in lithium-ion rocking-chair batteries with high charging capacities. Nanotube films embedded in electrolytes have been shown to be efficient electromechanical actuators, with possible application as artificial muscles. Nanotubes are also being considered as energy-storage and -delivery systems because of possibilities of hydrogen storage and excellent electron-transfer characteristics.

As Tromp writes, the nanometer world has properties and promises of its own that go beyond scaling and that may create entirely new technologies based on new physics, materials, and paradigms. The worldwide drive to invest in and develop nanotechnology, he believes, reflects the optimism that such a technology will soon become reality and that it will spawn new capabilities, opportunities, and wealth. Novel nanoscale materials—quantum dots, for instance—have applications that more conventional materials do not offer.[12]

Tromp cites nanocrystal semiconductor memory as an application based on the small capacitance of quantum dots (or nanocrystals, as they are often called in this context). In a conventional field-effect, he explains, transistor inversion is obtained by applying a suitable bias voltage to the gate. The incorporation of nanocrystals in the gate insulator would provide a means to apply a field offset by charging the quantum dots. Working devices have been successfully manufactured.

Tromp thinks that patterned media will be the next step in magnetic storage. As he explains, rather than use continuous magnetic thin films, researchers divide the film in spatially separated magnetic dots, where a single dot stores a single bit. This has certain advantages in terms of the ultimate storage density but is itself limited by the paramagnetic effect at a density of about 100 gigabits per square inch. When the bits become too small, the magnetic moment is subject to thermal fluctuations, and the stored information is subject to thermal decay. Recent attempts at fabricating magnetic nanocrystals have made much progress, and studies of their magnetic properties are under way. Hard magnetic thin films have also been fabricated using magnetic nanocrystals, even though the use of such nanocrystals in patterned magnetic media is not imminent.

---

[12]Frontiers of Engineering/2000. "Nanoscale Materials: Synthesis, Analysis and Applications," Rudolf M. Tromp, pp. 93-102.

Tromp also points out the benefits of nanostructural inks, which represent a generalization of the previous two applications. Manufactured in bulk, nanostructural inks can be applied to a wide variety of substrates using a wide variety of application methods, including inkjet printing, screen printing, spray application, spinning, and immersion. Inks can be mixed, layered, and patterned as applications require. Nanostructural inks might present the largest opportunity for the practical application of quantum dots and nanocrystals in the next two decades.

TREND 4: NATURE AS ENGINEERING TEMPLATE

Dickinson writes that organisms experimented with form and function for at least 3 billion years before the first human manipulations of stone, bone, and antler.[13] When pressed with an engineering problem, humans often draw guidance and inspiration from the natural world, using natural architectures as inspiration for so-called biomimetic, or bionic, design. Just as biologists are discovering the structural and physiological mechanisms that underlie the functional properties of plants and animals, engineers are beginning to develop means of fabrication that rely on nature for inspiration. As the performance gap between biological structures and mechanical analogs shortens, engineers may feel increasingly encouraged to seek and adopt such design concepts.

Dickinson points out that birds and bats, for instance, played a central role in one of the more triumphant feats of human engineering, that of airplane construction. In the 16th century, Leonardo da Vinci sketched designs for gliding and flapping machines based on anatomical study of birds. More than 300 years later, Otto Lilienthal built and flew gliding machines that were also patterned after birds. The wing-warping mechanism that enabled Orville and Wilbur Wright to steer their airplane past the cameras and into the history books is said to have been inspired by watching buzzards soar near their Ohio home.

Innovations in materials science, electrical engineering, chemistry and molecular genetics are enabling designers, Dickinson explains, to plan and construct complicated structures at the molecular or near-molecular level. Examples include buckyballs, nanotubes, and the myriad of microelectromechanical devices (MEMs) constructed with technology derived from the silicon chip industry. Integrated circuits themselves play a role in bionics projects aimed at constructing smart materials or mimicking the movement, behavior, and cognition of animals. Biological structures are complicated; only recently have engineers developed a sophisticated enough toolkit to mimic the salient features of that complexity. For

---

[13]Frontiers of Science/1999. Michael H. Dickinson, at <http://www.pnas.org/cgi/content/full/96/25/14208>.

their part, biologists are also beginning to understand how basilisk lizards walk on water, how penguins minimize drag, and how insects manage to remain airborne.

Dickinson writes that the fields of biology that use principles of structural engineering and fluid mechanics to draw structure/function relationships are known as functional morphology or biomechanics. These disciplines are of particular use to bionics engineers, he maintains, because the behavior and performance of natural structures can be characterized with methods and units that are directly applicable to mechanical analogues. In recent years, however, biomechanics has become increasingly sophisticated, aided by a battery of techniques, including x-ray cinematography, atomic-force microscopy, high-speed video, sonomicrometry, particle-image velocimetry, and finite-element analysis.

Dickinson cites a number of successful biomimetic designs that are based on the morphology of biological materials. A simple and well-known example is Velcro, invented by George de Mestral, who was inspired by the hours wasted pulling burrs off his dog's fur after walks in the Swiss countryside. In the same category is the lotus leaf: Although living above muddy water and without active grooming capability, lotus leaves remain pristine and dirt free. This self-cleaning ability, Dickenson notes, results from tiny, wax-coated protuberances on the lotus-leaf surface. When water falls on a leaf, it does not spread out and wet the surface, as it would on the smooth leaves of most plants, but rather forms tiny beads atop the knobby surface, collecting dust and dirt as they roll off. A brand of paints is now available that makes use of a patented "LotusEffekt" to form a similar protective barrier.

Another example is that of the shark. As do many fast-swimming organisms, sharks exhibit skin scales that possess tiny ridges running parallel to the longitudinal body axis. Dickinson points out that the grooved body surface reduces drag through its influence on the boundary layer between turbulence and smooth water flow. Placed over the wings and fuselage of an Airbus 320, riblet sheets modeled on sharkskin reduced the aircraft's fuel consumption.

Designers and engineers can mimic and utilize biological structures, Dickinson believes, provided that it is possible to fabricate the artificial material with the precision required to produce the desired effect. In the case of synthetic sharkskin, once engineers determined the correct groove geometry, it was relatively easy to mold plastic sheets that reproduce the pattern. House paints replicating lotus leaves are presumably laced with a material to mimic the rough surface of the leaves.

According to Dickinson, an example that well illustrates the crudeness of current microfabrication techniques is spider silk. Silks are proteins secreted by specialized glands found in many groups of arthropods. More than 4000 years ago, the Chinese domesticated the moth *Bombyx mori*, the primary source of textile silk. Although the quality of moth silk was great enough to have fueled the oldest intercontinental trade route in world history, its properties pale compared with those of spider silk. Spiders make a variety of different silks to serve different functions, but most research focuses on the dragline silk that individual spiders use to hoist and lower their bodies. This silk can extend and stretch by 30 percent without snapping; it is stronger than the best metal alloys or synthetic polymers. The use of ropes, parachutes, and bulletproof vests spun of spider silk has motivated the search for genes that encode silk proteins.

Dickinson cites the exoskeleton of insects as a good example of biological structural sophistication. He writes that the cuticle surrounding an insect is composed of one topologically continuous sheet composed of proteins, lipids, and the polysaccharide chitin. Before each molt, the cuticle is secreted by an underlying layer of epithelial cells. Complex interactions of genes and signaling molecules spatially regulate the exact composition, density, and orientation of proteins and chitin molecules during cuticle formation. Temporal regulation of protein synthesis and deposition permits construction of elaborate layered cuticles that display the toughness of composite materials.

The result of such precise spatial and temporal regulation is a complex exoskeleton that is tagmatized into functional zones. Limbs consist of tough, rigid tubes made of molecular plywood, connected by complex joints made of hard junctures separated by a rubbery membrane. The most elaborate example of an arthropod joint is the wing hinge, the morphological centerpiece of flight behavior. The hinge consists of a complex interconnected tangle of five hard elements embedded within thinner, more elastic cuticle and bordered by the thick side walls of the thorax. In most insects, the muscles that power the wings are not attached to the hinge. Instead, flight muscles cause small strains within the walls of the thorax, which the hinge then amplifies into large oscillations of the wing. Small control muscles attached directly to the hinge enable the insect to alter wing motion during steering maneuvers.

Although the material properties of the elements within the hinge are indeed remarkable, Dickinson asserts, it is the structural complexity as much as the material properties that endow the wing hinge with its unique characteristics. Several research groups are actively attempting to construct miniature flying devices patterned after insects. Their challenge is not simply to replicate an insect wing, Dickinson notes, but to create a mechanism that flaps it just as effectively.

Microscale Materials

Chakraborty believes that certain classes of plastics—polymers—could be designed from the molecular scale up in order to perform microscale functions.[14] Nature uses proteins and nucleic acids in much the same way, he writes: Evolu-

---

[14]Frontiers of Engineering/1999. "Design of Biomimetic Polymeric Materials," Arup K. Chakraborty, pp. 37-43.

tion has devised schemes that allow the design of macromolecular building blocks that can self-assemble into functionally interesting structures. This does not mean that researchers should precisely copy the detailed chemistries of nature. Rather, they are exploring underlying universality in the schemes that natural systems employ to carry out a class of functions and are determining whether these adaptations affect biomimetic behavior.

According to Chakraborty, many biological processes, such as transmembrane signaling and pathogen-host interactions, are initiated by a protein when it recognizes a specific pattern of binding sites on part of a membrane or cell surface. Recognition means that the polymer quickly finds and then adsorbs strongly on the pattern-matched region and not on others. He writes that the development of synthetic systems that can mimic such recognition between polymers and surfaces could have a significant impact on such advanced applications as the development of sensors, molecular-scale separation processes, and synthetic viral-inhibition agents. Chakraborty believes the ability of certain classes of molecules to form organized assemblies in solution has important commercial and biological consequences. A variety of products, such as detergents, emulsifiers, catalysts, and vehicles for drug delivery, already rely on this ability.

TREND 5: THE MATURATION OF AUTONOMOUS MACHINES

The field of smart materials and structures combines the knowledge of physics, mathematics, chemistry, computer science, and material, electrical, and mechanical engineering in order to accomplish such tasks as making a safer car, a more comfortable airplane, and structures capable of self-repair, assert Cao et al.[15] The trio write that, in the future, with the help of miniaturized electromechanical devices, structures may be "intelligent" enough to communicate directly with the human brain. The development of supersensitive noses, ears, and eyes would enable humans to smell more scents, hear beyond a certain frequency range, and see what normally cannot be seen naturally, such as the infrared spectrum.

A smart structure, explain the three scientists, is a system containing multifunctional parts that can perform sensing, control, and actuation; it is a primitive analogue of a biological body. Smart materials are used to construct these smart structures, which can perform both sensing and actuation functions. The "I.Q." of smart materials is measured in terms of their responsiveness to environmental stimuli and their agility. The first criterion requires a large amplitude change, whereas the second assigns faster response materials a higher I.Q.

---

[15]Frontiers of Science/1998. Wenwu Cao, Harley H. Cudney, and Rainer Waser, at <http://www.pnas.org/cgi/content/full/96/15/8330>.

Commonly encountered smart materials and structures can be categorized into three different levels, according to Cao et al.: single-phase materials, composite materials, and smart structures. Many ferric materials and those with one or more large anomalies associated with phase-transition phenomena belong to the first category. Functional composites are generally designed to use nonfunctional materials to enhance functional materials or to combine several functional materials to make a multifunctional composite. The third category is an integration of sensors, actuators, and a control system that mimics the biological body in performing many desirable functions, such as synchronization with environmental changes and self-repair of damage. A smarter structure would develop an optimized control algorithm that could guide the actuators to perform required functions after sensing changes.

Active damping is one of the most studied areas using smart structures assert Cao et al. By using collocated actuator and sensors (i.e., physically located at the same place and energetically conjugated, such as force and displacement), a number of active damping schemes with guaranteed stability have been developed. These schemes are categorized on the basis of feedback type in the control procedure, i.e., velocity, displacement, or acceleration.

Goldberg writes that economic pressures and researcher interest continue to press the evolution of simple control structures into more complex, semi-intelligent directions. Although anticipated for perhaps two decades, the widespread distribution of robots and products with some robotic capability—most notably, systems that monitor or direct the manufacture of goods—are becoming routine. While household robots nevertheless remain the stuff of science fiction, microprocessor capability continues to advance sufficiently to raise the prospect of nontoy robotic devices in households within the decade.[16]

Goldberg estimates that by 1998, there were 700,000 robots at work in industry. Almost half those robots were installed in Japan, while 10,000 were installed in the United States and Germany, respectively, and the remaining 20,000 were installed in Korea, Italy, France, and other countries in the developed world. By far the largest application areas are welding and painting, followed by machining and assembly. The largest customers are the automotive industry, followed by electronics, food, and pharmaceuticals. Robotics also continues to be an active and thriving area of research, with ongoing studies of kinematics (positions and velocities), dynamics (forces), and motion planning (to avoid obstacles).

Because motion planning is designed to allow robots to reach their destination without being stymied by difficult or unanticipated obstacles, robotics research is finding an unexpected benefit, according to Kavraki. She writes that pharmaceutical firms are using the results to plan the routes of therapeutic

---

[16]Frontiers of Engineering/1998. "A Brief History of Robotics," Kenneth Y. Goldberg, pp. 87-89.

molecules to their docking sites on a variety of bodily proteins, in order to treat or cure disease.[17]

Collaborative robots, or "cobots," are a new type of robotic device, writes Michael Peshkin, intended for direct interaction with a human operator within a shared workspace. Cobots allow a true sharing of control between human and computer. The human operator supplies motive power and exerts forces directly on the payload, while the mechanism of the cobot serves to redirect or "steer" the motion of the payload under computer control. The computer monitors the force (direction as well as magnitude) applied by the operator to the payload.

In real time, Peshkin notes, these operator forces can be compared with programmed guiding surfaces, and motion in the direction that the operator pushes can be allowed, disallowed, or redirected. The human operator may be allowed complete freedom of motion of the payload, or in the opposite extreme, the payload may be restricted to a single curve through space. Thus, the full versatility of software is available for the production of virtual surfaces and other sensate effects.[18]

Cobots rely on the worker to provide motive power, or can give some powered assistance that requires only small motors. The much greater need for force is that required for changes of direction, sometimes called "inertia management." In cobots this is accomplished by the physical mechanism of the cobot rather than by motors, with a consequent improvement in both safety and smoothness of operation.

Peshkin believes that cobotic applications beyond manufacturing are several. In image-guided surgery, for example, safety is essential; a cobot's ability to guide motion without possessing a corresponding ability to move on its own can alleviate concern about inadvertent movement due to software or hardware malfunction. Because the quality of a virtual surface enforced by a cobot originates in its physical mechanism rather than in servocontrolled actuators, harder and smoother surfaces are possible than can be achieved by a conventional robot. Preserving the critical sense of touch in surgery requires a high-quality shared control between surgeon and robot, for which smoothness of motion is essential.

Popular weight training equipment, originally designed for rehabilitation, uses shaped cams and other mechanical components to confine a user's limb or body motion to a particular trajectory. Peshkin observes that while these trajectories are somewhat adjustable, far greater versatility could be achieved if the motion trajectories were encoded in software rather than frozen into the mechanical design of the equipment. Cobots can enforce virtual trajectories with appropriate levels of smoothness, hardness, and safety.

---

[17]Frontiers of Engineering/1998. "Algorithms in Robotics: The Motion-Planning Perspective," Lydia E. Kavraki, pp. 90-94.

[18]Frontiers of Engineering/1998. "Cobots," Michael A. Peshkin, pp. 105-109.

At present, conventional robots' interface to computers and information systems is their primary benefit. Collaborative robotics allows for comingling computer power with the innate intelligence of humans and the full responsiveness of their senses and dexterity. Thus far, such capabilities cannot be matched or replaced by robots alone.

ADDITIONAL TOPICS OF NOTE

Certain fields, while not trends in and of themselves, represent studies into disciplines that are intriguing and potentially relevant. A limited discussion of six such areas follows. The sections were taken from the original papers and edited for clarity. The language is primarily, and the assertions completely, those of the cited authors.

Climate Change

Studies of past climate changes show that the Earth system has experienced greater and more rapid change over larger areas than was generally believed possible, jumping between fundamentally different modes of operation in as little as a few years. Ongoing research cannot exclude the possibility that natural or human-caused changes will trigger another oscillation in the near future.[19]

Global climate change is of interest because of the likelihood that it will affect the ease with which humans make a living and, perhaps, the carrying capacity of the planet for humans and other species. Attention is focused on the possibility that human activities will cause global climate change, because choices affect outcomes.

Long climate-change records show alternations between warm and cold conditions over hundreds of millions of years associated with continental drift. Large glaciers and ice sheets require polar landmasses, continental rearrangement, and associated changes in topography that affect oceanic and atmospheric circulation; in turn, these affect and are affected by global biogeochemical cycles, which include high levels of atmospheric carbon dioxide, associated with warm times.

According to Alley et al., the last few million years have been generally cold and icy compared with the previous hundred million years but have alternated between warmer and colder conditions. These alternations have been linked to changes over tens of thousands of years in the seasonal and latitudinal distribution of sunlight on Earth caused by features of the Earth's orbit. Globally synchronous climate change—despite some hemispheric asynchrony—is explained

---

[19]Frontiers of Science/1998. Richard B. Alley, Jean Lynch-Stieglitz, and Jeffrey P. Severinghaus, at <http://www.pnas.org/cgi/content/full/96/18/9987>.

at least in part by the lowering of carbon dioxide during colder times in response to changes in ocean chemistry. We currently live in one of the warmer, or "high," times of these orbital cycles. Previously, the coolest "low" times brought glaciation to nearly one-third of the modern land area.

Recent examination of high-time-resolution records has shown that much of the climate variability occurred with spacings of one to a few thousand years. Changes within high times have been large, widespread (hemispheric to global, with cold, dry, and windy conditions typically observed together), and rapid (over periods as short as a single year to decades).

The changes have been especially large in the North Atlantic basin. In the modern climate, the warm and salty surface waters of the Gulf Stream heat the overlying atmosphere during winter, becoming dense enough to sink into and flow through the deep ocean before upwelling and returning in a millennium or so. Numerical models and paleoclimatic data agree that changes in this "conveyor belt" circulation can explain at least much of the observed millennial variability, although the reconstructed changes may be more dramatic than those modeled. Sinking can be stopped by North Atlantic freshening associated with increased precipitation or with melting of ice sheets on land and a resulting surge into the ocean. North Atlantic sinking also might be stopped by changes in the tropical ocean or elsewhere.

Of concern to Alley et al. is that some global warming models project North Atlantic freshening and possible collapse of this conveyor circulation, perhaps with attendant large, rapid climate changes. At least one model indicates that slowing the rate of greenhouse gas emissions might stabilize the modern circulation.

Fluorescence Sensing

After a long induction period, say de Silva et al., fluorescent molecular sensors are showing several signs of wide-ranging development.[20] The clarification of the underlying photophysics, the discovery of several biocompatible systems, and the demonstration of their usefulness in cellular environments are key indicators. Another sign is that the beneficiaries of the field are multiplying and have come to include medical diagnostics through physiological imaging, biochemical investigations, environmental monitoring, chemical analysis, and aeronautical engineering.

The design of fluorescent molecular sensors for chemical species combines a receptor and a fluorophore for a "catch-and-tell" operation. The receptor module engages exclusively in transactions of chemical species, while the fluorophore is

---

[20]Frontiers of Science/1998. A. Prasanna de Silva, Jens Eilers, and Gregor Zlokarnik, at <http://www.pnas.org/cgi/content/full/96/15/8336>.

concerned solely with photon emission and absorption. Molecular light emission is particularly appealing for sensing purposes owing to its near-ultimate detectability, off/on switchability, and very high spatiotemporal resolution, including video imaging.

Neurobiology, say the authors, has been a special beneficiary of fluorescence-sensing techniques combined with high-resolution microscopy. This is so because the central nervous system is a highly complex network of billions of cells, each with an elaborate morphology. Enhanced understanding of the central nervous system requires more intense study of neuronal activity at the network level as well as increased subcellular resolution.

Neotectonics

Recent advances in Global Positioning System (GPS) technology have made it possible to detect millimeter-scale changes in Earth's surface. According to Clement et al., these advances have made it possible to detect relative motion between the large plates of the outermost rigid layer of Earth.[21] These motions previously had only been inferred from indirect evidence of the plates' motions. A remarkable piece of knowledge from these studies is that the plate motions are nearly continuous, not episodic, processes, even on human time scales. Analyses of these motions indicate that much of the motion between plates occurs without producing earthquakes. In addition to monitoring interplate motions, GPS arrays are making it possible to study present deformation occurring within mountain belts. The strain-rate models obtained from the GPS arrays then can be used to test specific theories of crustal deformation in mountain belts.

The theory of plate tectonics that revolutionized the earth sciences during the 1960s was based primarily on indirect evidence of past crustal movements. Motions of the seafloor crust were inferred from the magnetization of the crust that recorded polarity reversals (of known age) of Earth's magnetic field. The symmetric magnetic patterns suggested that new crust was being created at the mid-ocean ridges and then was spreading away from the ridges. The seafloor-spreading hypothesis was successful in explaining many longstanding problems in earth sciences and became the basis of a new paradigm of crustal mobility.

Most plate boundary zones accommodate motion along relatively narrow regions of deformation. However, plate boundary zones involving continental lithosphere absorb relative motion by deforming over broad zones that are hundreds to even thousands of kilometers wide. An understanding of the forces at work in these zones is important because many of the most damaging earthquakes occur within these zones, say Clement et al. At present, little is known of how the

---

[21]Frontiers of Science/1999. Bradford Clement, Rob McCaffrey, and William Holt, at <http://www.pnas.org/cgi/content/full/96/25/14205>.

relative motion between two tectonic plates is accommodated by earthquakes and how much is taken up by slow creep, either steady or episodic. Understanding the ratio of fast, seismic (earthquake-producing) slip to slow aseismic slip is fundamentally important in the quest to assess the danger of active geologic faults.

The use of GPS arrays capable of continuously monitoring a large region provides the resolution needed to monitor short- and long-term displacements that occur during and after earthquakes. In addition to making estimates of the component of aseismic creep between major earthquakes, we also can now estimate the relative amounts of seismic and aseismic slip associated with a particular earthquake. Cases have been documented in which the aseismic slip after an earthquake has accommodated as much slip or more slip than the quake itself. If this is a common occurrence, more than half of plate-boundary slip may be aseismic.

Several GPS arrays are being deployed across plate boundaries in an effort to monitor slip events. For example, an array being installed across the Cascadia subduction zone offshore of Oregon and Washington State is capable of detecting purely creep events. These events result from slip on a fault that does not radiate seismic energy detectable by seismometers and hence do not produce traditional earthquakes. These creep events may explain the paradox that many fault zones currently have high strain rates, although their histories are largely devoid of earthquakes, or the quakes that did happen were too small to account for the long-term rate of slip.

Extrasolar Planets

The discovery of extrasolar planets has brought with it a number of surprises. To put matters in context, say Najita et al., the planet Jupiter has been a benchmark in planet searches because it is the most massive planet in the solar system and is the object that we are most likely to detect in other systems.[22] Even so, this is a challenging task. All the known extrasolar planets have been discovered through high-resolution stellar spectroscopy, which measures the line-of-sight reflex motion of the star in response to the gravitational pull of the planet. In our solar system, Jupiter induces in the Sun a reflex motion of only about 12 meters per second, which is challenging to measure given that the typical spectral resolution employed is approximately several kilometers per second. Fully aware of this difficulty, planet-searching groups have worked hard to achieve this velocity resolution by reducing the systematic effects in their experimental method.

---

[22]Frontiers of Science/1999. Joan Najita, Willy Benz, and Artie Hatzes, at <http://www.pnas.org/cgi/content/full/96/25/14197>.

As one example, prior to detection, the stellar light is passed through an iodine gas-filled absorption cell to imprint a velocity reference on the stellar spectrum. However, after search techniques had been honed in this way for years to detect a Jupiter-like world in other solar systems, a surprising result has emerged: a much greater diversity of planetary systems than was expected. According to Najita et al., searches have revealed planets with a wide range of masses, including planets much more massive than Jupiter; planets with a wide range of orbital distances, including planets much closer to their suns than Jupiter is to our Sun; and planets with a wide range of eccentricities, including some with much more eccentric orbits than those of the planets in our solar system.

These results were essentially unanticipated by theory and reveal the diversity of possible outcomes of the planet-formation process, an important fact that was not apparent from the single example of our own solar system. This diversity is believed to result from the intricate interplay among the many physical processes that govern the formation and evolution of planetary systems, processes such as grain sticking and planetesimal accumulation, runaway gas accretion, gap formation, disk-driven eccentricity changes, orbital migration, and dynamical scattering with other planets, companion stars, or passing stars. Thus far, what has changed is not so much our understanding of the relevant physical processes but, rather, how these processes fit together, i.e., our understanding of their relative importance and role in the eventual outcome of the planet formation process.

Femtochemistry

The essence of the chemical industry and indeed of life is the making and breaking of molecular bonds. The elementary steps in bond making and breaking occur on the time scale of molecular vibrations and rotations, the fastest period of which is 10 femtoseconds. Chemical reactions are, therefore, ultrafast processes, and the study of these elementary chemical steps has been termed "femtochemistry." According to Tanimura et al., a primary aim of this field is to develop an understanding of chemical reaction pathways at the molecular level.[23] With such information, one can better conceive of new methods to control the outcome of a chemical reaction. Because chemical reaction pathways for all but the simplest of reactions are complex, this field poses both theoretical and experimental challenges. Nevertheless, much progress is being made, and systems as complex as biomolecules can now be investigated in great detail.

Ultrafast dynamics of molecules have long been studied theoretically by integrating a relevant equation of motion. The time-dependent wave packet ap-

---

[23]Frontiers of Science/1998. Yoshitaka Tanimura, Koichi Yamashita, and Philip A. Anfinrud, at <http://www.pnas.org/cgi/content/full/96/16/8823>.

proach has proven to be particularly promising for following femtosecond chemical reactions in real time. Briefly, a molecular system can be characterized by the electronic potential energy of surfaces on which wave packets propagate.

Experimental efforts in the field of femtochemistry have exploited the pump-probe technique, wherein a pump laser pulse initiates a chemical reaction and a probe laser pulse records a "snapshot" of the chemical reaction at a time controlled by the temporal delay between the pump and probe pulses. By recording snapshots as a function of the temporal delay, one can follow the time evolution of a chemical reaction with time resolution limited only by the duration of the laser pulses. Beyond monitoring the outcome of a normal photoreaction, the phase and frequency of a femtosecond pump pulse can be tailored, as prescribed by theory, to drive a molecular state to a target location on its potential energy surface and then steer it toward a channel that favors a particular photochemical outcome.

For example, say the authors, the excitation pulse might be a femtosecond, linear-chirped laser pulse, which can interact with the wave packet through a so-called intrapulse, pump-dump process. A negatively chirped pulse (frequency components shift in time from blue to red) might be tailored to maintain resonance with the wave packet as it evolves along the excited state surface. In contrast, a positively chirped pulse might quickly go off resonance with the wave packet, and the photoexcitation would be nonselective.

Understanding molecular motions and how they couple to the reaction coordinate is crucial for a comprehensive description of the underlying microscopic processes. This problem is particularly challenging because molecules exhibit strong mutual interactions, and these interactions evolve on the femtosecond time scale because of random thermal motion of the molecules. In essence, understanding the dynamics of a molecular system in the condensed phase boils down to a problem of nonequilibrium statistical physics. Combined with an impressive increase in computational capacity, recent developments in theoretical methodology such as molecular dynamics, path-integral approaches, and kinetic-equation approaches for dissipative systems have enlarged dramatically the scope of what is now theoretically tractable.

Biometric Identification

Personal identification, regardless of method, is ubiquitous in our daily lives. For example, we often have to prove our identity to gain access to a bank account, to enter a protected site, to draw cash from an ATM, to log in to a computer, to claim welfare benefits, to cross national borders, and so on. Conventionally, we identify ourselves and gain access by physically carrying passports, keys, badges, tokens, and access cards, or by remembering passwords, secret codes, and personal identification numbers (PINs). Unfortunately, passports, keys, badges, tokens, and access cards can be lost, duplicated, stolen or forgotten, and passwords, secret codes, and PINs can easily be forgotten, compromised, shared, or observed.

Such loopholes or deficiencies of conventional personal identification techniques have caused major problems for all concerned, say Shen and Tan.[24] For example, hackers often disrupt computer networks; credit card fraud is estimated at $2 billion per year worldwide; and in the United State, welfare fraud is believed to be in excess of $4 billion a year. Robust, reliable, and foolproof personal identification solutions must be sought to address the deficiencies of conventional techniques.

At the frontier of such solutions is biometrics-based personal identification: personal physical or biological measurements unique to an individual. Some frequently used measurements are height, weight, hair color, eye color, and skin color. As one may easily observe, although such measurements can accurately describe an individual, more than one individual could fit the description. To identify an individual based on biometric data, the data should be unique to that individual, easily obtainable, time-invariant (no significant changes over a period of time), easily transmittable, able to be acquired as nonintrusively as possible, and distinguishable by humans without much special training. The last characteristic is helpful for manual intervention, when deemed necessary, after an automated, biometrics-based identification/verification system has made an initial determination.

Automated biometrics-based personal identification systems can be classified into two main categories: identification and verification. In a process of verification (one-to-one comparison), the biometrics information of an individual who claims a certain identity is compared with the biometrics on the record for the person whose identity is being claimed. The result of the comparison determines whether the identity claim is accepted or rejected. On the other hand, it is often desirable to be able to discover the origin of certain biometrics information to prove or disprove the association of that information with a certain individual. This process is commonly known as identification (one-to-many comparison).

According to Shen and Tan, a typical automated biometrics-based identification/verification system consists of six major components. The first is a data-acquisition component that acquires the biometric data in digital format by using a sensor. For fingerprints, the sensor is typically a scanner; for voice data, the sensor is a microphone; for face pictures and iris images, the sensor is typically a camera. The quality of the sensor has a significant impact on the accuracy of the comparison results. The second and third components of the system are optional. They are the data compression and decompression mechanisms, which are designed to meet the data transmission and storage requirements of the system. The

---

[24]Frontiers of Science/1998. Weicheng Shen and Tieniu Tan, at <http://www.pnas.org/cgi/content/full/96/20/11065>.

fourth component is of great importance: the feature-extraction algorithm. The feature-extraction algorithm produces a feature vector, in which the components are numerical characterizations of the underlying biometrics. The feature vectors are designed to characterize the underlying biometrics so that biometric data collected from one individual at different times are similar, while those collected from different individuals are dissimilar. In general, the larger the size of a feature vector (without much redundancy), the higher its discrimination power. The discrimination power is the difference between a pair of feature vectors representing two different individuals. The fifth component of the system is the "matcher," which compares feature vectors obtained from the feature extraction algorithm to produce a similarity score. This score indicates the degree of similarity between a pair of biometrics data under consideration. The sixth component of the system is a decision maker.

H

Trends in Science and Technology

Patrick Young

CONTENTS

INTRODUCTION	168
TRENDS IN INFORMATION TECHNOLOGY	168
Computers, Lithography, and Thin Films, 168	
Data Density, 170	
Flat Displays and Printed Circuitry, 172	
The Internet, 174	
Imaging, 175	
TRENDS IN MATERIALS SCIENCE AND TECHNOLOGY	177
Micro- and Nanoscale Fabrication, 177	
Photonics, 179	
Superconductors, 180	
TRENDS IN ENERGY, ENVIRONMENT, AND THE PLANET	183
Energy, 184	
Environment and Ecology, 185	
Atmospheric Sciences and Climatology, 187	
Earthquake Studies, 188	
TRENDS IN BIOMEDICAL AND AGRICULTURAL SCIENCES	189
Genomics and Proteomics, 190	
Neuroscience, 191	
Drug Development, 193	
Agrobiotechnology, 194	
BLENDING THE PHYSICAL AND THE BIOLOGICAL	195
Biotech and Materials Science, 195	

Robot Engineering, 196
Catalysts, 196
DISCUSSION AND CONCLUSION 197

INTRODUCTION

Any attempt to predict the future carries risks. Yet, in forecasting the scope of technology advances over a span of 5 to 10 years, the past is prologue to the future. During the first decade of the 21st century, three enabling technologies—computers, communications, and electronics—will continue and accelerate the information revolution. These technologies, in turn, will be supported by innovations across a spectrum of supporting technologies, such as advanced imaging, nanotechnology, photonics, and materials science. Important developments will emerge as well in such areas as sensors, energy, biomedicine, biotechnology, and the interaction of the physical and biological sciences.

Many challenges, however, confront those seeking to turn the potential of these technologies into practical applications. How rapidly progress comes will depend on innovative ideas, the economy, and the vision of industrial and political leaders.

TRENDS IN INFORMATION TECHNOLOGY

In about 10 years, the semiconductor industry will require a replacement technology for the photolithography process now used to make electronic chips if it wants to keep improving their performance. Two advanced fabrication approaches are under investigation at a time when the industry is also in transition from the use of aluminum to copper for circuit lines and is seeking ways to further shrink the size of transistors. Computer makers expect to vastly increase data storage over the next decade, using new magnetic and optical techniques that include magnetic thin films, near-field optics, holography, and spintronics. Scientists will also continue the quest for higher-resolution and flatter display panels, pursuing such approaches as organic light-emitting diodes and electronic "paper." New and refined imaging techniques will enable cutting-edge studies in the physical and biological sciences.

Computers, Lithography, and Thin Films

Chipmakers must confront the time, probably toward the end of the current decade, when traditional lithography techniques no longer meet their needs. The Semiconductor Industry Association's periodic technology road map makes clear that the decade ahead will require unprecedented changes in the design of chips and their materials if the industry is to continue its traditional rate of doubling computing power about every 18 months. Potential approaches to improved performance include a new chip fabrication technology, improved circuitry, and new types of transistors.

Shrinking the size of microcircuits, the hallmark of the semiconductor industry, has relied in considerable part on the reduction in the light wavelength used for optical lithography. Line widths of 200 nanometers are now in use, but by the end of this decade or soon after, lithography as we know it will likely reach its physical limits. The industry is now pursuing two advanced lithography techniques as potential replacements for the current technology. Each of the two—extreme ultraviolet light (EUV) and projection electron beam lithography (also known as SCALPEL, for scattering with angular limitation projection electron-beam lithography)—has its advantages and limitations. EUV provides radiation at 13.4 nanometers and uses a complex mirror system rather than lenses for focusing it on the photoresist. The process, however, will require defect-free masks—an extraordinarily difficult challenge—and, most likely, some type of chemical post-processing of the photoresist material to ensure proper depth of the etching. SCALPEL forsakes photons for high-energy electrons, which poses problems in developing masks that are both thin enough for the exposure process and capable of withstanding its high heat. Solutions to this problem exist, but they require complex processing to achieve the finished chip.

Faster Speeds

In an effort to ensure faster computing speeds, chipmaking is in a fundamental transition as it moves away from the historic combination of aluminum circuit lines coated with an insulator, or dielectric, usually silicon dioxide. Following the lead of IBM, companies are substituting copper for aluminum. The choice of an improved dielectric—needed to prevent cross talk as line width continues to narrow—is far less unanimous. IBM has chosen Dow Chemical's SiLK aromatic hydrocarbon, but other companies, including Novellus Systems and Dow Corning, offer competing dielectrics, and other new insulating materials are under investigation.

Innovative ways to shrink transistors remain high on the research agenda. One approach would reduce the thickness of the gate insulator, which is currently about 2 nanometers. Achieving chips with 100-nanometer lines will likely require shrinking the insulator to 1 nanometer, only four times the width of a silicon atom. Several other potential options for reducing insulator width are open. One is to reduce the gate length, which would increase speed without having to shrink the insulator layers. This might be accomplished by silicon-on-insulator transistors, in which the insulator is placed under the transistor rather than burying the transistor in the silicon substrate. Another approach, the double-gate transistor, would place one gate atop another and reduce the gate length by half for the same oxide thickness.

The switch to copper wires required new advances in plasma processing of microcircuits, which, in turn, allowed chipmakers a wider selection of materials. Further improvements will be necessary, however, for generating such things as diamond thin films for use in flat-panel displays. Improvements in the deposition of organics and thin films are vital to improving the performance of the next generation of electronic devices. Today, the material to be deposited restricts the choice of technology used. A more universal deposition technique would offer a significant advantage.

Nonvolatile RAMs

Recent years have seen a surge of interest in developing inexpensive, fast, durable, and nonvolatile random access memories and in the use of a solid-state technology called magneto-electronics to replace volatile and nonvolatile semiconductor memories and mechanical storage. Most of this work is focused on giant magnetoresistance (GMR) and magnetic tunnel junction technology. GMR materials have the advantage of a strong signal, nonvolatility, and compatibility with integrated circuit technology. Magnetic tunnel junction devices have the potential to serve as nonvolatile memories with speeds comparable to those of today's dynamic random access memories (DRAMs). Much of the research on magneto-electronic memories has emphasized hybrid devices that utilize a magnetic memory and semiconductor electronics. But one start-up company has gone all-metal, developing a magnetic RAM in which the memory arrays and the electronics are all made of GMR materials. The device is based on electron spin rather than electric charge.

Data Density

More bits per square inch is almost a mantra in the computer data-storage field. But as with photolithography, traditional magnetic and optical storage techniques are approaching their own physical limits, which will require innovative solutions to overcome. In 1999, IBM forecast that it would achieve an areal magnetic storage density of 40 gigabits per square inch (Gb/in.^{2}) by the middle of this decade. Beyond this density, magnetic storage encounters the instability of superparamagnetism, a phenomenon in which the magnetic orientation energy equals the surrounding thermal energy. As a result, magnetic bits flip spontaneously at normal operating temperatures. In May 2001, IBM announced it had developed a new coating for its hard disk drives that bypasses the problem—a three-atom-thick layer of ruthenium sandwiched between two layers of magnetic material. IBM predicted that the material would enable a storage density of 100 Gb/in.^{2} by 2003.

Optical storage faces its own physical barrier—the diffraction limit, at which the size of the optical bits is limited by the wavelength of light used to record them. Solving the two problems will require new materials, structures, and recording technologies.

Magnetic Storage

Today's magnetic storage devices, such as hard disk drives, record data on tiny tracks of cobalt-chromium alloy crystals. One approach being explored to increase density in these devices involves creating a thick film of copolymer plastic, burning holes in it as small as 13 nanometers, and filling them with magnetic materials. Because this technique could yield 12 trillion magnetic "posts" or "wires" over a single square centimeter, each separated from the others by plastic, it could result in magnetic storage significantly higher than the 40-gigabit limit imposed by superparamagnetism on conventional systems. Indeed, it might someday boost data density into the terabit range.

Another way to improve density would be to replace the cobalt-chromium crystals used in storage media with iron-platinum particles, which have stronger magnetism and could be made as small as 3 nanometers. However, until last year, when IBM scientists succeeded, no one could produce uniform grains of the metal crystals. Uniform grains with greater magnetic strength should enable data densities up to 150 Gb/in.^{2} and, perhaps, even into the terabit range.

Both approaches, however, face a number of challenging development and scaling issues before they are ready for the market, and they will require new read and record technologies as well if they are to reach their full potentials.

Optical Storage

Near-field optics—which exploits the fact that light placed very near an aperture smaller than its wavelength can pass through the hole—may provide one way to vastly expand the data density of optical storage. Using a very small aperture laser, Lucent scientists have recorded and read out optical data at a density of 7.5 Gb/in.^{2}, and they have speculated that with apertures 30 nanometers in diameter, data density could reach 500 Gb/in.^{2} Lucent has licensed its technology for commercial development.

Holography offers the potential for high storage densities and data transfer at billions of bits per second for two reasons. Unlike traditional magnetic and optical systems, holography can store data throughout the entire medium rather than simply on the surface. Second, holography allows recording and reading out a million bits at once rather than one bit at a time.

For example, InPhase Technologies, a Lucent spin-off, is commercializing technology developed at Bell Laboratories. It uses two overlapping beams of light—the signal beam, which carries data, and the reference beam. The two beams enter the storage medium at different angles, create an optical interference pattern that changes the medium's physical properties and refractive index, and are recorded as diffractive volume gratings, which enables readout of the stored data.

Encoding data consists of assembling "pages" of 1 million bits represented by 1's and 0's and sending them electronically to a spatial light modulator. This device is coated with pixels, each about 10 square micrometers, which can be switched rapidly to match the content of each page. When a signal beam passes through the modulator, its pixels either block or pass light, depending on whether they are set as a 1 or a 0, and the laser beam carries the message of that specific page.

Bringing competitive holographic systems to market that will exceed the traditional storage technologies will require a number of advances in recording materials. These advances include optical clarity, photosensitivity, dimensional stability, and uniform optical thickness, as well as innovations in spatial light modulators, micromirrors, and component-systems integration.

The demonstration that information can be stored on and nondestructively read from nanoclusters of only two to six silver atoms, announced earlier this year by Georgia Institute of Technology researchers, opens another potential approach to increasing data density. The Georgia Tech team exposed a thin film of the silver nanoclusters to blue light in the shape of the letter L. Two days later, they exposed the nanoclusters to green light, which caused the nanoclusters to fluoresce in the L pattern. Whether such nanoclusters can be shaped into compact arrays and handle read-write operations at the speeds of today's computers remains a question for further study.

Electron Spin

Spintronics could lead to information storage on the same chips that process data, which would speed up computation. Data processing is based on the charge carried by electrons; data storage has relied on magnetism or optics. However, electrons also have spin, and electron spin is harnessed in magnetic storage. Spintronics seeks to manipulate electron spin in semiconductor materials for data storage and perhaps quantum computing. The key lies in devising semiconductor materials in which spin polarized electrons will function. Recent developments in spin polarizers and the synthesizing of magnetic semiconductors suggest this problem can be managed. However, making a marketable product will require ferromagnetic semiconductors that operate at room temperature—a demand not easily fulfilled.

Flat Displays and Printed Circuitry

Organic light-emitting diodes (OLEDs), a technology that offers more design flexibility and higher resolution than traditional LEDs, are now coming to market. However, these devices are limited in size, so they are not yet practical for such things as monitors and television screens.

OLEDs rely on small-molecule oligomers or thin films of larger semiconductor polymers for their illumination. A polymer semiconductor, for example, is deposited on a substrate, inserted between electrodes, and injected with electrons and holes (the absence of electrons). When holes and electrons recombine, they emit light. The technology is expected to one day replace cathode ray tubes and liquid-crystal diodes. Advocates emphasize several advantages of light-emitting polymer-based products, including greater clarity, flatter screens, undistorted viewing from greater angles, higher brightness, and low drive voltages and current densities, which conserve energy. OLEDs for alphanumerical use and backlit units for liquid crystal diodes are currently entering the marketplace.

The coming decade will probably see innovations in OLED production, materials, performance, and scale. Advances in all these areas are needed to bring to market such envisioned products as high-sensitivity chemical sensors, roll-up television screens, wide-area displays, and plastic lasers.

Electronic Paper

This technology may change the configuration and the way we use portable electronic devices such as cell phones and laptops as well as reinvent how newspapers, books, and magazines are "printed" and read. The vision is of a lightweight, rugged, flexible, and durable plastic that combines the best of wood-pulp-based paper and flat-panel displays. In a sense, the earliest versions of the vision are available today. E Ink Corp. markets a simple version of electronic paper for large-area displays. Gyricon Media, Inc., a Xerox Corp. spin-off, plans to market a precursor electronic paper for similar uses later this year.

The design of electronic paper differs markedly from the electronic displays of today. Instead of cathode-ray tubes or liquid-crystal diodes, silicon circuits, and glass, electronic paper would utilize electronic "inks," plastic "paper," and flexible and bendable circuitry. A joint venture by Lucent Technologies and E Ink unveiled the prototype last fall, a device containing 256 transistors, each of which controls a single pixel. A thin layer of ink made of white particles suspended in a black fluid is placed between two electrodes. Switching a pixel on or off causes the white particles to move forward or backward and make the pixel appear black or white.

To form an electronic paper, inks must be laminated along with their drive circuitry into flexible sheets. This poses a problem because plastics typically cannot withstand the high temperatures needed for manufacturing conventional silicon circuits. Moreover, the surfaces of plastics are rougher than those of glass or silicon, which can adversely affect viewing. So making electronic paper a viable commercial product will require a number of technological developments.

Electronic paper and innumerable other products would benefit from the ability to simply print electronic circuits rather than go through the stressful and complex process used to make chips. The goal is to fabricate transistors, resistors, and capacitors as thin-film semiconductor devices. Ways to do this at low cost and in large volume using standard printing processes or inkjet printers are in development. Working with funds from the Advanced Technology Program, for example, Motorola has teamed with Dow Chemical and Xerox in a 4-year effort to develop novel organic materials and techniques for printing electronic devices.

The Internet

Rarely, if ever, has a technology changed society as rapidly and unexpectedly as the Internet and the World Wide Web did in the 1990s. The coming decade will also see rapid changes in the way the world communicates and transmits data.

The Internet was both revolutionary and evolutionary, a dual process that continues with the next-generation Internet, or Internet II. Initiated at a conference in October 1995, Internet II involves a collaboration of more than 180 universities and a multitude of federal agencies and private companies to develop an advanced communications infrastructure. The major goals are to create a cutting-edge network for the research and education communities, enable new Internet applications, and ensure that new services and applications get transferred to Internet users at large. Designers envision high-speed, low-loss, broadband networks capable of allowing such bit-dense activities as real-time research collaborations and telemedicine consultations of unsurpassed clarity. Internet II encompasses several major new Internet protocols, and, as the original Net did, it will introduce new phrases to the language, such as GigaPOP—the term used for its interconnection points between users and the providers of various services. Among the many innovations needed to enable Internet II are new network architectures, advanced packet data switch/routers, multiplexers, and security and authentication systems.

Internet Vulnerability

The growth of the Internet has stimulated the study of communications networks to understand their general properties and the physical laws that govern their behavior. Physicists at the University of Notre Dame did a computer simulation of two possible network configurations. In one, each node had about the same number of connections to other nodes in the network. In the second configuration, nodes had greatly varying numbers of connections but nodes with many connections predominated. The second configuration represented the type of connections found on the Internet. On the basis of their findings, the researchers concluded that Internet-like systems are largely invulnerable to random failure but very open to damage by deliberate attack. A better understanding of the

Internet's behavior and its effect on Internet vulnerability is important for national security, communications within and among businesses, and the flow of e-mail and e-commerce.

The introduction of the Advanced Encryption Standard algorithm last year could assure that information encrypted by it and sent over the Internet cannot be decoded by anyone who intercepts it, at least for the next several decades. The challenge now is to ensure the security of the encryption process so that the specific information needed to decode messages remains known only to those who should know it. This is primarily an issue of people and policy. However, NIST is working on techniques called key-management protocols that will help enforce the security of the encryption and decoding process.

A major challenge, one that could have a significant impact on Internet reliability and speed from the user's viewpoint, lies in resolving the so-called last-mile bottleneck. This is the connection between the desktop terminal and the Internet service provider. Technical advances in optical communications, some of them associated with developing Internet II, will shorten this last mile stretch by stretch, increase network communications, and perhaps even solve the bottleneck in the coming decade.

Imaging

Imaging has served as a vital impetus to discovery across the spectrum of science for several centuries. This fact will remain true in the 21st century.

Advances in imaging at the nano- and molecular scales by various techniques have contributed significantly to the understanding and exploitation of materials and processes. As science seeks to understand and control nature at its smallest scales, the need for new and improved imaging techniques—more sensitive, more specific, sharper in detail—takes on new urgency. New approaches and refinements of old, reliable methods will certainly emerge in the coming decade. Femtosecond lasers, for example, are opening a new era of investigation, ranging from biochemical reactions to fundamental studies of quantum mechanics. Optical microscopy, the oldest of the imaging sciences, and imaging holography could find new uses. Improvement in synchrotron-radiation resolution promises sharper images of such things as chemical-bond orientation, individual magnetic domains, solid-state reactions, catalysts, and the surfaces of semiconductors.

Femtosecond Imaging

Refinements in femtosecond imaging and its application to new areas promise greater understanding in a broad range of disciplines, from cell biology to materials science. The laser-based technique already has demonstrated, for example, that DNA is not a rigid molecule but is capable of considerable motion. Currently, laser pulses of 5 femtoseconds can be achieved. The discovery that fast x-ray pulses can be generated when ultrashort laser pulses are reflected from the boundary between a vacuum and a plasma suggests the potential for a new femtosecond imaging approach. Researchers envision opportunities such as observing the biochemical reaction of new drugs, precisely defining the transport of electrons within DNA, and even gaining a greater understanding of quantum theory through the application of femtosecond imagery.

Light Microscopy

The original microscopy has yet to reach its limits of usefulness, especially in areas such as biotechnology, biomedical science, and medical diagnostics. By integrating advances from several fields, including optics, robotics, and biochemistry, researchers are developing interactive light microscopy techniques to examine the contents and dynamics of cells and tissues. For example, the National Science Foundation is funding development of the automated interactive microscope, which couples advanced fluorescence-based light microscopy, image processing, and pattern recognition to a supercomputer. Researchers are also exploring deblurring techniques to sharpen the images yielded by innovative light microscopes.

Near-field scanning optical microscopy takes advantage of the fact that light shined through a tiny nanoaperture can cast an illumination spot the size of the hole. Spot sizes in the range of 50 to 20 nanometers have been obtained. German researchers have reported using a single molecule as a light source. In theory, such a light source could illuminate a spot approximately 1 nanometer across.

One current approach to developing nondestructive imaging on the nanoscale combines two established technologies—scanning probe microscopy and molecular spectroscopy. The aim is to harness the high spatial resolution offered by scanning probes and molecular spectroscopy's chemical specificity to explore the chemical details of nanometer structures. Holography, too, offers a potential means of imaging at the nanoscale. Working with three partners and money from the Advanced Technology Program, nLine Corp. seeks to develop a holographic system capable of imaging defects on the bottoms of deep, narrow features of semiconductor chips. These features include trenches, contacts, and the spaces between interconnects, where depth-to-width ratios run as high as 30 to 1.

Algorithms

In many instances, the key to improved imaging will be new algorithms and software packages, through which, for example, researchers can obtain enhanced image quality, create extremely accurate three-dimensional images from the observations of different devices, automate image correction, and gain more interactive capabilities with images. Synchrotron radiation facilities of increased intensity and more precise focus will advance observations in protein structure, surface science, and molecular structure. New variations and refinements of scanning probe and atomic force microscopy can be expected to improve imaging of the physical and biological worlds, helping to solve issues in biochemistry and nanostructure synthesis and fabrication, and to advance the quest for biomolecular devices such as high-density processors, optical-communications elements, and high-density storage media.

TRENDS IN MATERIALS SCIENCE AND TECHNOLOGY

Fabrication at the micro- and nanoscale level will yield a number of new devices and products, ranging from exquisite sensors to tiny walking robots and automated labs-on-a-chip. Key to such advances is understanding how to control the materials used and the development of new molecular manipulation and micromachining tools. Advancements in photonics will help meet the demand for greater bandwidth for communications, and innovations in photonic-integrated systems will expand their use for signal processing. Both high- and low-temperature superconductors pose challenges and promise commercial applications over the next decade, including in the distribution of electric power and as ships' engines. Creation of new materials will have effects throughout society, and the versatility of polymers makes them a particularly attractive target for research. Self-assembly, by which molecules form into structures on their own, also has gained increasing attention.

Micro- and Nanoscale Fabrication

Emerging micro- and nanominiaturization techniques promise to transform the typically planar world of these scales into three dimensions and to enable new devices and technologies of scientific, industrial, economic, and security import, including a host of new sensors, walking microrobots, nanomotors, and new polymers. Understanding how to control the chemical composition, physical properties, and configuration of materials at the molecular level is a key element in devising nanoscale building blocks for assembly into working devices and machines. Achieving this knowledge and integrating it into products requires an interdisciplinary effort by chemists, physicists, materials scientists, and engineers.

MEMS

Microelectromechanical system (MEMS) devices have gone from relatively simple accelerometers to devices that enabled the successful flight of twin tethered experimental communications satellites, each of which is only 12 cubic inches and weighs 0.55 lb. The challenge now is to improve MEMS techniques and develop new ways to do three-dimensional microfabrication of things such as metallic coils. MEMS devices—already in commercial use as sensors and actuators—are being developed as micromachines and microrobots. Especially promising to the MEMS field is the advent of laser-based micromachining and other innovations to supplement the photolithography-chemical etching process used originally. Laser techniques now in laboratory development can micromachine metals and fabricate devices of smaller dimensions—a few micrometers today and, perhaps, nanoscale devices tomorrow. Swedish researchers have fabricated a walking silicon microrobot, as well as a microrobotic arm that uses conjugated-polymer activators to pick up, move, and place micrometer-size objects.

At the nanoscale level, the ability to manipulate atoms, molecules, and molecular clusters has revealed often unexpected and potentially useful properties. Now scientists are seeking to exploit these findings to create nanowires, nanomotors, macromolecule machines, and other devices. Nanomotors, for example, will be needed to power many envisioned nanodevices, including switches, actuators, and pumps. The challenge is to devise ways to convert chemical energy into power that enables nanodevices to perform useful tasks and to find ways to control and refuel the tiny motors.

Nanotubes

Carbon-based nanotubes continue to attract interest because of the potential of their physical and electrical properties. They are stronger and tougher than steel, can carry higher current densities than copper, and can be either metals or semiconductors. Uses envisioned for them range from tiny switches to new composite materials capable of stopping high-velocity bullets. Bringing such nanometer applications to fruition, however, especially for electronics, will require new methods to reliably and economically mass-produce nanotubes and control their characteristics, as well as ways to structure and organize them.

Without scanning probe microscopy, nanotechnology would remain largely a concept. Nevertheless, there exists the need for faster techniques to control, image, and manipulate materials and for new ways to make molecular-scale measurements. One approach to greater specificity would harness the spatial resolution of scanning probe microscopy with the chemical specificity of molecular spectroscopy.

Sensors

Microfabrication and MEMS devices will play increasingly greater roles in the development and manufacture of high-tech sensors. The increased potential for terrorist attacks and the threat of chemical or germ warfare has spurred efforts by the civilian and military sectors to detect explosives, deadly chemicals, and biological agents. A single lab-on-a-chip can detect and identify a number of difference substances, a capability useful to medical and environmental monitoring as well as national security. Beyond today's technology, researchers funded by the Defense Advanced Research Projects Agency (DARPA) are developing small, lightweight, easy-to-use MEMS-based biofluidic microprocessors capable of monitoring a person's blood and interstitial fluid and comparing readings with an assay reference chip. Such devices could not only give early warning of exposure to chemicals or biological agents but also monitor general health, medication usage, and psychological stress.

The coming decade should see other significant advances in medical microsensors. For example, English scientists are developing a camera-in-a-pill that can be swallowed to examine the gastrointestinal tract. The device, currently 11 × 30 millimeters, contains an image sensor, a light-emitting diode, telemetry transmitter, and battery. Several improvements are needed before it can complement or replace current endoscopy tools, including orientation control and a more powerful battery that will enable it to image the entire gastrointestinal tract, from ingestion to elimination. And a Michigan company, a winner in the 2000 Advanced Technology Program competition, is trying to commercialize technologies developed at the University of Michigan. It hopes to create implantable wireless, batteryless pressure sensors to continuously monitor fluids in the body. Potential beneficiaries include patients with glaucoma, hydrocephalus, chronic heart disease, or urinary incontinence.

Artificial or electronic noses are starting to find a home in industry. These devices typically consist of an array of polymers that react with specific odors and a pattern-recognition system to identify an odor by the change it produces in a polymer. To date, artificial noses have been used mainly in the food industry to augment or replace existing means of quality control, but potential uses include quality control in the pharmaceutical, cosmetic, and fragrance industries. Expanding the uses of electronic noses will require sensors with greater sensitivity and specificity and more advanced algorithms to improve performance.

Photonics

Photonics underpins optical communications. Because photons are more effective carriers of information than electrons, the demand for new ways to harness light to communicate and store data will intensify, driven by the worldwide need for greater bandwidth. The development of wavelength division multiplexing has opened a progressive revolution in data transmission, storage, and processing that could match that of electronics in the 20th century. Although most photonic circuits today are analog, the development of low-cost photonic integrated systems will enable uses beyond communications, including signal processing and exquisite sensors.

Photons travel at the speed of light and do not interact with one another, which all but eliminates cross talk and inference. However, logic functions require some interaction, and researchers have sought ways to process information electronically and transmit it optically. One promising approach to integration is the use of smart pixel devices, which combine electronic processing circuitry with optical inputs and/or outputs and can be integrated into two-dimensional arrays. Researchers are investigating several approaches to applying smart pixels in high-speed switching, as interconnects, and in flat-panel display applications.

Electro-optical Polymers

The increase in optical-fiber capacity also creates the need to speed information in and out of the fibers. One new approach to speeding this information flow uses polymers containing organic chromophores, which are molecules involved in color. Evidence suggests that embedding chromophores can yield electro-optical polymers that have higher speeds—as high as 100 gigahertz—and require lower voltages than present-day electronic modulators. However, chromophores tend to align in ways that reduce the effect of an applied field, a problem that needs resolution.

Researchers are also exploring holography to create three-dimensional photonic crystals for use in ultrasmall waveguides and other optical-communications uses. So far, however, the thickness of the crystal layers is limited to about 30 micrometers, and more advances in processing will be needed to obtain the larger photonic crystals needed for communications applications.

Significant improvements in key photonic devices seem a certainty in the next decade, including in-fiber optical filters (known as fiber Bragg gratings), in-fiber amplifiers, and fiber lasers. Fiber Bragg gratings, for example, are the building blocks of wavelength division multiplexing and essential elements for the next generation of all-optical switches and networks. In-fiber amplifiers depend on doping the core of optical fibers with rare-earth ions, and refining the doping process should optimize various amplifier designs. Fiber lasers can sustain high power densities and can currently generate pulses of 100 femtoseconds. The development of polymer lasers holds significant potential for speeding communications.

One challenge to developing new photonic devices is that models currently used to characterize materials require unrealistic computer time. As a result, prototype devices often must be built for testing. A great need exists for new algorithms to address issues such as simulation, analysis, alignment tolerance, and increasing the yields of photonic components.

Superconductors

In recent years, applications of high-temperature superconductors have moved beyond the realm of laboratory sensors. The U.S. Navy, for example, has awarded a contract for the design of a 25,000-horsepower superconductor motor to power its next generation of destroyers. The ability of these superconductors to transmit electricity with essentially zero resistance will in and of itself guarantee continued efforts to understand the phenomenon, develop new superconducting materials, and apply them.

Opening up applications and a large commercial market for superconductor wires and cables will require increasing the current-carrying capacity of high-temperature superconductors, and the coming years will witness further efforts to resolve this challenge. German researchers have achieved a sixfold increase in current-carrying capacity at 77 K by manipulating the make-up of a yttrium-barium-copper oxide superconductor. The researchers replaced some yttrium ions at grain boundaries with calcium ions. 77 K is the boiling point of nitrogen and the point above which many believe superconductors can be widely commercialized. This discovery suggests similar chemical tinkering may raise the current-carrying capacity of other high-temperature superconductor compounds.

Superconducting Plastics

Electrically conducting plastics were discovered a quarter-century ago, but only now have superconductor plastics come to the fore. Superconductor polymers have been problematic because plastics carry current as the result of doping them with impurities, but superconductivity requires an ordered structure that does not disrupt electron flow. An international team that included Bell Labs researchers reported earlier this year that it had observed superconductivity in the polymer poly(hexythiophene). The researchers placed the polymer inside a field-effect transistor, injected it with holes, and achieved superconductivity at 2.35 K. The new discovery should pave the way for a new approach to developing and applying superconductor polymers.

The discovery of an intermetallic superconductor with a critical temperature of 39 K, announced earlier this year, opens another area of opportunity. Although low in operating temperature compared with today's high-temperature superconductors, the magnesium boride compound can carry three times as much current, weight for weight, because it is a lighter material, and it can be cooled by closed-cycle refrigeration instead of liquid helium coolants.

Nanotubes

Low-temperature superconductors continue to demonstrate resilience and the potential for significant development. An ongoing Department of Energy program supports efforts to move these materials into commercial use. In one demonstration project, a consortium of companies is readying superconductor cable to serve 14,000 inner-city Detroit customers with electricity. Experimentally, nanotubes—which at room temperatures can be insulators, semiconductors, or metals—have added to their reputation as physical chameleons with the discovery that they can become superconductors. European researchers suspended a number of single-wall nanotubes between two superconductor electrodes and found they became superconductors and ultrasmall Josephson junctions at between 0.3 and 1.0 K, depending on the sample.

New Materials and Self-Assembly

Materials science underpinned much of the technological advancement of the 20th century and will continue to do so in this decade. The discipline involves understanding the structure and physical properties of materials and finding ways to alter them for useful purposes. The coming years will accelerate the design of materials with specific performance capabilities dictated by a specific need. The applications of materials science range across modern life, from new structural materials, to sensing devices, to implantable biocompatible replacement parts for humans. As they have in recent decades, innovations in materials will enable a spectrum of new products and technologies.

Polymers

These long-chain molecules have made important contributions to technology and will continue to do so (see Flat Displays and Printed Circuits). Polymers, depending on their composition, can be as rigid and strong as steel, as floppy and stretchable as rubber bands, or somewhere in between. Synthesizing polymers with specific properties, sizes, and shapes, however, remains a significant challenge. Despite substantial progress in controlling the architecture and functions of polymers, considerably more needs to be accomplished in understanding these materials and in developing ways to easily tailor their configurations and properties to meet specific needs—whether as new materials for an old product or a new material for an envisioned use. Japanese researchers, for example, have used the probe tip of a scanning tunneling microscope to fashion conjugated polymers into nanowires. Researchers at the Massachusetts Institute of Technology are developing a shape-memory polymer with properties and applications superior to those of the shape-memory metal alloys now available.

Composite Materials

By chance, guesswork, and science, humans have created alloys for more than 4000 years. Composite materials, in the form of straw-based bricks, go back roughly 7000 years. As scientists develop a deeper understanding of how to structure and process molecules in ways that fine-tune their physical properties, new alloys and composite materials will emerge. In Earth-orbit experiments, researchers have developed a nanocomposite in which magnetic nanoparticles move freely inside cavities that form within the material. Potential applications include tiny compasses, gyroscopes, switches, and, perhaps, microtransformers. Today's alchemists rely on computers—for modeling, computational chemistry, and computational physics—sensitive imaging studies, and nondestructive testing. Even better tools are needed as the science surges forward.

Self-Assembly

One area likely to progress rapidly over the next decade is that of molecular self-assembly to inexpensively produce atomically precise materials and devices. Harnessing this phenomenon will provide a critical element in fabricating new materials, nanomachines, and electronic devices. Some examples of self-assembling materials in development include smart plastics that assemble themselves into photonic crystals, spinach-based opto-electronic circuits for possible use in logic devices and ultrafast switches, and inorganics and organics that self-assemble between two electrodes to form transistors. Earlier this year, English scientists reported what they called the equivalent of catalytic antibodies for synthetic chemists—a dynamic solution that enables molecules to arrange themselves into the best combination to bind to a specific target. Self-assembly has important applications in the biomedical sciences. One technique, for example, uses a system in which molecules assemble into countless combinations that are quickly tested for their ability to bind to receptors.

A great deal remains unknown about the process of molecular self-assembly or how to utilize it for producing new materials or new applications. Even more challenging is the creation of materials that self-assemble from two or more types of molecules.

TRENDS IN ENERGY, ENVIRONMENT, AND THE PLANET

Fuel cells will enter the marketplace shortly as part of the power systems of automobiles and for use as portable electric generators, and they may soon provide power for smaller devices such as cell phones and laptop computers. Solar cells and nuclear power plants will both receive greater attention in the next decade, but significant problems remain to be solved for both. Hormone disrupters, environmental pollutants, global warming, and ozone disintegration are and will remain issues of public and international concern, as will ecosystems and their preservation. Studies should also reveal whether large-scale sequestering of carbon dioxide is practical and more about long-term patterns of climate change and shorter-term episodes, such as El Niño. Although precise earthquake prediction will remain but a goal, geoscientists should gain a greater understanding of the mechanics and behavior of quakes, and this knowledge should inform the earthquake engineering of buildings and other structures.

Energy

Energy is again a national issue, and the Bush administration's energy policy, while emphasizing increased production, includes interest in conservation, efficiency, and innovation. The century's first decade will likely see increased research related to nuclear power and ways to store electricity in large quantities, but the technologies most likely to have significant impact in the energy area exist today. One can predict with near-certainty that no magic bullet will emerge to cure the nation's energy ills.

Fuel Cells

Discovered in 1839, fuel cells long remained little more than a potential source of clean and efficient energy. Today, they are poised to become a commercial commodity. The 1990s saw a resurgent interest in fuel-cell technology, largely in proton-exchange membrane (PEM) fuel cells for use as nonpolluting vehicle power plants. PEM fuel cells split hydrogen atoms into electrons and protons and produce water and heat. The first production-model fuel-cell cars should be available in 2004, and some automakers expect to be selling 100,000 fuel-cell-equipped vehicles a year by 2010. The viability of fuel cells for vehicles resulted from a number of technological advances that improved fuel processing at lower costs, and further improvements will come. Current automotive fuel cells convert the hydrogen in fossil fuels such as methanol to electrical energy. Ultimately, the makers of PEM fuel cells expect to replace hydrocarbons with pure hydrogen. Two potential energy sources for producing the needed hydrogen from water are high-efficiency solar cells and nuclear power plants.

Solid oxide fuel cells (SOFCs) are in development as sources of electric power, and at least one company expects to market a model for home use later this year. SOFCs essentially work in the reverse manner of PEMs. They add electrons to oxygen, and the oxygen ions then react with a hydrocarbon to produce electricity, water, and carbon dioxide. Their potential applications include supplying electricity to individual buildings, groups of buildings, or entire neighborhoods. Conventional SOFCs operate at about 1000 °C, which causes materials stress and poor stability. Efforts are under way to find ways to improve material reliability and reduce operating temperatures.

Solar Cells

Although solar cells have a small niche in today's energy production, their low conversion rate of sunlight to electricity—about 20 percent at best—remains a stumbling block to their greater use. Significant advances in solar-cell materials are urgently needed. One company, Advanced Research Development, has combined two polymers, polyvinyl alcohol and polyacetylene, to produce solar cells that it contends could reach a conversion rate at which nearly 75 percent of solar energy becomes electricity. German researchers have invented a photoelectrolysis cell that they believe can, with improvements, convert 30 percent of solar energy into electricity to produce hydrogen.

Nuclear

The renewed interest in fission-driven power plants preceded the Bush administration, but the administration's policies and the current shortages of electricity will add impetus to the relicensing of some existing plants and perhaps interest in building new plants before the end of the decade. However, although the nuclear industry has developed new designs for power plants, issues of materials, safety measures, radioactive-waste disposal, and the negative public view of nuclear power remain formidable barriers to bringing new plants online. The revival of interest in nuclear plants, however, suggests the need to pursue work in these areas, particularly ways to improve disposal techniques for nuclear plant wastes, because some states limit on-site storage within their boundaries, and the federal government has yet to open a national storage site.

Environment and Ecology

Environmental problems did not start with the industrial age, but industrialization did exacerbate them. Understanding and countering environmental threats to human health and Earth's flora, fauna, air, and water will challenge researchers across many disciplines in the coming decade. Progress in environmental science will depend, in part, on new technologies that enable faster evaluation of chemicals, advances in computational biology, new animals models, better databases, and defining the interaction of environmental factors and genes.

Environmental Pollutants

One growing concern is the question of hormone disrupters, or hormone mimics, which are environmental chemicals that evidence suggests may interact with the endocrine systems of humans and animals to cause birth defects and several cancers. A key element in investigating the issue is the need to develop reliable, short-term assays to identify hormone-disrupting chemicals.

Occupational exposure, safe water, and metal contaminants also remain a concern. Although many workers nationwide encounter chemical and biological agents in the course of their employment, the level and risk of exposure for the vast majority remains unknown or poorly defined. Pollutants contaminating water supplies and the use of chlorine pose unresolved scientific and policy questions. "Safe" levels of such contaminants in water as arsenic and the impact of the by-products of chlorination on human illness remain unknown. Metals in the environment—especially lead, mercury, arsenic, cadmium, and chromium—pose health threats and problems in the remediation of industrial and hazardous waste sites. New approaches to removing contaminants are needed, and a number of them, including biosurfactants, should reach the market in the next few years.

Genomics' Role

The mapping of the human genome will play a significant role in understanding how environmental factors affect human health and the genetic susceptibility of people to various chemical pollutants and infectious agents. The first challenge is to identify genes and/or specific polymorphisms that interact with these agents, as well as metabolic, nutritional, and behavioral factors. These genes include those that influence metabolism, detoxification, cell receptors, DNA repair, immune and inflammatory responses, oxidation pathways, and signal transduction. From this effort should come ways to prevent and intervene in disease processes, both at the individual and population level.

Concerns about pollution in general are contributing to green chemistry, which is the development of environmentally safer chemical processes. One area likely to see a quiet revolution is the development of new solvents that are less volatile and therefore less likely to reach the atmosphere and do harm. Ionic fluids, which have a low vapor pressure at room temperatures, are one example. Another approach seeks to alter conventional solvents in such ways that they retain their desirable activity yet are environmentally benign or less harmful.

Ecosystems

As important as ecosystems are to food production, sequestering carbon, salinity control, and biodiversity, they remain poorly understood. The public recognition of their importance, however, has led to efforts to restore some ecosystems to their original vitality and to some frustration. The history of a specific ecosystem is often lost through the system's years of alteration, and this loss of knowledge includes the interactions of such factors as water flow and the species of plant life that once flourished there. Moreover, attempts to restore traditional grasses, shrubs, and trees to lure back specific wildlife may fail because of a lack of understanding about what is essential for a successful interaction between the two. If ecosystem restorations are to succeed as intended, many new details of their complexity must be uncovered.

The harm from importing alien species has been documented for more than a century: the disappearance of the American chestnut, the spread of Dutch elm disease, the invasion of the zebra mussel, and other economically disastrous ecological events. Attempts to block the arrival of alien species also require ways to eliminate or control those that establish themselves. Achieving success in this effort seems unlikely in the next 5 to 10 years.

Atmospheric Sciences and Climatology

Humankind's influence on the atmosphere and the world's climate will gain further attention worldwide this century. The seasonal Antarctic ozone hole may have reached a natural limit, but the Arctic region—where ozone thinning occurs but to a lesser extent than over the southern polar region—causes a greater concern because it has a far larger population potentially susceptible to the harmful effects of the Sun's ultraviolet radiation. New evidence suggests that 10- to 20-micrometer particles of frozen water, nitric acid, and sulfuric acid form in the stratosphere in winter water over the Arctic. These particles—known as polar stratospheric cloud (PSC) particles—have perhaps 3000 times the mass of previously known PSC particles and provide a surface on which chlorine and bromine convert from inactive to active molecules that destroy ozone. This discovery illustrates the many unknown effects that gases and submicrometer particles released into the atmosphere—and their interactions—can have. Probing these questions will require not just more measurements and more sensitive sampling and analytical techniques but also new, testable theories that apply atmospheric chemistry, fluid mechanics, and thermodynamics to the problem. Aerosol particles pose a particular challenge because unlike gases, which can remain in the atmosphere for decades, particles in the micrometer range typically last but a few days.

Global Warming

The vast majority of scientists who have assessed the issue now regard global warming, and the human contribution to it through the burning of fossil fuels, as real. This view was reinforced in a recent report by the National Academy of Sciences/National Research Council to President Bush.[1] Climate simulations reported last year, for example, compared two sets of data: natural contributions to global warming, such as volcanic activity and variations in solar radiation alone, versus natural contributions plus those from human activities. The results indicated that had natural factors alone prevailed, atmospheric warming, in the 20th century would have stopped about 1960 and a cooling trend would have followed. Combining human and natural activities, however, produced a continued warming pattern up to the present.

Aside from reducing the burning of fossil fuels, several other approaches are under investigation to limit the release of carbon dioxide into the atmosphere. One of these would inject carbon dioxide produced by large combustion plants underground into salt-brine-containing rock to sequester it. Another possibility, but one that requires far more study, is to dissolve carbon dioxide in the ocean

---

[1] National Research Council, Committee on the Science of Climate Change, Climate Change Science: An Analysis of Some Key Questions, 2001.

depths. A major challenge to sequestering the greenhouse gas is to find ways to reduce the cost of such storage to about $10 a ton, a goal the Department of Energy has set for 2015.

Climate Patterns

Sorting out how to best inhibit the harmful effects of human activities on the atmosphere will take on a new urgency in the coming years. The problem has important international economic and social implications—such as coastal submersion from rising sea levels, increased storms and flooding, and disruptions in agriculture—that reinforce the need for excellent science. The Bush administration's emphasis on the use of fossil fuels for electricity generation and combustion engines does not bode well for reducing greenhouse gases in the near term.

Studies of ice cores from Greenland and the Antarctic continue to elucidate evidence that Earth goes through periodic temperature shifts short of full ice ages. Recent data indicate that this warming–cooling cycle occurs poles apart: When it grows colder in the Arctic, the southern polar region warms and vice versa. Work correlating rises and falls in Earth's temperature with ocean circulation patterns may further explain the intricate interconnection that shifts the climate on scales of hundreds and thousands of years.

Climate changes of shorter scales, such as the phenomena known as El Niño and La Niña, present a challenge of even greater human immediacy. El Niño is a movement of warm surface water from the western Pacific to the eastern Pacific off South America. It is propelled by the Southern Oscillation, an unstable interaction between the ocean and the atmosphere. During an El Niño, the trade winds weaken and the warm water rushes east and releases more water into the air, which results in heavy rainfall over Peru and Ecuador. Far to the west, places such as Indonesia and Australia suffer droughts. La Niña is the reverse phase of the oscillation. So expansive is the area of warm Pacific water that it affects climate over much of the globe in both phases. The two strongest El Niños in more than 100 years occurred in 1982 and 1997, and some scientists believe this intensification was a result of global warming.

Earthquake Studies

The introduction and refinement of plate tectonics provided a unifying theory that in the broadest terms explained the pattern of earthquakes and volcanic activity observed globally, such as the so-called ring of fire around the rim of the Pacific Ocean. Yet knowing generally where a quake might strike does not say when, and plate tectonics itself did little to provide the information needed to accurately predict temblors by precise time and location.

Earthquakes occur when stress builds within rock, primarily along faults beneath the surface, to the point the rock breaks and moves. The coming decade, with near certainty, will not yield ways to pinpoint earthquakes in time and space. However, one can predict a better understanding of the buildup of stress, the transfer of stress along faults, and the complex ways the shockwaves are released when rock snaps propagate through the lithosphere and along the surface. This information will have implications for emergency planning and for improving the design and construction of earthquake-resistant structures.

Earthquake Behavior

Geoscientists in recent years have advanced their understanding of earthquake behavior in several ways. For example, the notion of fault-to-fault communication has shown that faults pass stress quite effectively at the time of a quake from one segment to another. During an earthquake, considerable stress is released as shockwaves, but some of the fault's stress also is transferred to adjacent segments. By determining the epicenter of a quake and the direction in which the rock broke—strike slip or thrust—seismologists can calculate how much stress was shifted to adjoining rock and where. This ability allows an informed estimate of where the fault will break next, but the question of when remains unanswered.

A better determination of timing could emerge from the new ability—made possible by more powerful computers—to pinpoint and observe clusters of thousands of microquakes along a fault over time. Seismologists believe these microquakes, which are magnitude 1 to 3, represent the progressive failure of a fault. However, what triggers the fault to finally break and release its pent-up energy also remains unknown. Currently, several teams are observing clusters of microquakes on the San Andreas fault near Parkfield, California, where a moderate earthquake has been expected to occur for well over a decade.

TRENDS IN BIOMEDICAL AND AGRICULTURAL SCIENCES

Sequencing the genomes of humans and other species, coupled with proteomics and advances in bioinformatics, will reveal the genes related to diseases, alter the way physicians practice medicine, and have a major impact on the development of new drugs. The growing understanding of the complex activity inside cells will provide equally important insights, as witnessed by research in the neurosciences. Among the emerging findings is that brain cells may be capable of regenerating themselves, and there is a better understanding of protein-protein interactions, such as those of hormones and their receptors. Biotechnology will play an increasingly important role in developing human drugs, but public resistance in places to genetically modified plants may slow its role in agriculture.

Genomics and Proteomics

The mapping and sequencing of the human genome, now nearing completion, marks a historic point in biology and the beginning of equally exciting discoveries as scientists make increasing sense of the jumbled A's, T's, G's, and C's (adenine, thymine, cytosine, and guanine) that make up its genes. This effort will see scientists reporting important data gleaned from the genome, such as the role of specific genes, the identification of genes that are linked to diseases, and a better understanding of the timing of gene expression. These efforts will require new or improved assay systems, automated processing equipment, computer software, and advances in computational biology.

Other Genomes

Of considerable importance to understanding the human genome are the continuing efforts to sequence the genomes of other creatures, including the mouse. Many genes are conserved; that is, the same gene exists in many species, and the functions of certain of these genes have been discovered in species other than humans. By matching, say, the mouse and human genomes, a gene with a known function in the mouse can be pinpointed in humans. During the next 7 years, progress along the frontlines of human genomics should identify most genes associated with various diseases and indicate how a malfunctioning gene relates to the ailment, which would open new windows to therapy.

Genomics will change many current approaches in medicine and related health sciences. One area likely to see radical change is toxicology. The field has traditionally relied on animals—such as rats, mice, rabbits, dogs—to gauge the toxicity of substances. But genomics research has led to an emerging field known as toxicogenomics. In it, researchers apply a suspected toxin to DNA placed on glass to observe any effects it may have on gene expression.

Proteomics

Genes carry the codes for proteins, which do the actual work within an organism. The genomic revolution has ushered in the age of proteomics, an even more difficult challenge in which the goal is nothing less than understanding the makeup, function, and interactions of the body's cellular proteins. Indeed, proteomics poses a more complex puzzle than genomics because there are far more proteins than genes. This is so because the messenger RNA that transports the code from a gene for transcription into a protein can be assembled in several ways, and a protein in a cell also can be modified by such processes as phosphorylation and glycosylation.

Proteomics, unlike the traditional study of one protein at a time, seeks to pursue its goals using automated, high-throughput techniques. The development of new technologies is a major goal. Areas of investigation include which genes the different cell types express and their proteins, the characteristics of these proteins, studies to determine their interactions, understanding signal transduction, and determining the nature of protein folding and the exact three-dimensional structure of proteins. As a rule, proteins that interact tend to work together, such as antibodies and their receptors. Identifying a protein active in some disease process offers a potential target for intervention. For this reason, the proteomics quest will be led and dominated by the quest to discover new drugs.

A key element in the success of genomics and proteomics has been and will continue to be bioinformatics, a field that in the broadest sense weds information technology and biology. Bioinformatics provides the computer systems and strategies needed to organize and mine databases, examine the relationship of polymorphisms to disease, ascertain the role of proteins and how they interact with other proteins, and inform the search for screening tests and therapeutic drugs. Once basically a data management tool, bioinformatics is now important in the analytical functions needed to support continuing progress in genomics and proteomics—not simply cataloging and warehousing data, but turning it into useful insights and discoveries. All this will require the development of new algorithms and computer strategies.

Neuroscience

Neuroscience provides but one example of the opportunities and advances that will accrue through a greater understanding of the extraordinarily complex activities within and among cells. The last three decades have brought a burst of knowledge about the brain—the routes of neurons throughout the organ, an understanding of neurotransmitters, synapses, and receptors, mappings of responses to various stimuli, and the linking of genes to biochemical networks, brain circuitry, and behavior. Physicians can now track the progress of neurological diseases in their patients with imaging techniques, and new insights into the processes of memory, learning, and emotions have emerged. One can expect the brain to yield far more of its secrets in the coming decade with the introduction of new probes, new imaging technologies, and improved bioinformatics techniques that integrate findings not only from neuroscience but from other research disciplines as well.

Nerve Death

Consider, as an example, neurodegeneration, which is a key element in ailments such as Alzheimer's, Parkinson's, and Huntington's diseases, multiple sclerosis, glaucoma, and Creutzfeldt-Jakob disease and its variant, commonly called mad cow disease. Evidence suggests a genetic component for each of these diseases, different degenerative mechanisms for each, yet with some similarities among them that may yield insight into more than one disease. In Alzheimer's, for instance, a genetic defect in the nucleus affects the cell's mitochondria, its major source of energy, and appears to play a role in triggering apoptosis, or programmed cell death. The question now being raised is whether the same or a similar mechanism may play a role in the optic-nerve degeneration in glaucoma.

The ability to regenerate central nervous system neurons could provide a major advance in treating neurodegenerative diseases and paralyzing spinal cord injuries. One target is the biochemical cascade of apoptosis, which ultimately ends in the death of a cell. Preliminary evidence now suggests that it is possible to interrupt the cascade at several points and that doing so may stop the process of cell death and perhaps return the cell to normal functioning.

Another target of opportunity is stem cells, partially developed cells that transform into the various cells of the body. Stem cells taken from embryos and fetuses and transplanted into patients once appeared to be the only stem-cell approach capable of replacing dead neurons. However, ethical and moral questions have slowed and delayed investigations of the uses of embryonic and fetal stem cells. Recently, several animal studies indicated that stem cells from adults can be reprogrammed to form specific tissues, including, perhaps, neurons. Studies to date are essentially observations and many key questions remain. What mechanisms determine what type of cell a stem cell will become? What stem cells enter the brain and become neuronlike? What signals attract them? Scientists will devote long hours during the next few years to deciphering these biological codes and seeking to utilize the answers in therapy.

Self-Repair

The brain itself might one day be stimulated to self-repair. A maxim for years held that central nervous system neurons did not and could not regenerate. Experiments during the last several years in animals, including mice, birds, and primates, have challenged that basic assumption. Some evidence suggests that apoptosis can, in certain circumstances, stimulate stem cells in the brain to form new neurons, and these neurons form the same connections with other neurons as the dead cells. Adult stem cells have also been stimulated to form new heart muscle in mice.

Although the 1990s were proclaimed the decade of the brain, the potential for advances in neuroscience during this decade is even greater. Improved techniques for imaging and mapping the biochemical pathways of the brain, more powerful bioinformatics tools and more interaction among scientists working in different areas of neuroscience should reveal considerably more information about the brain. One can expect a better understanding of the neurodegenerative diseases, the functioning and interactions of dendrites, synapses, and ribisomes, the mechanisms of neurotransmitter production and transport, internal cellular signaling, and learning and memory.

Drug Development

Biotechnology allows the transfer of new genes into animals and humans, the culturing of plants from single cells, and the development of new drugs and diagnostic tests. Knowledge gleaned from the human genome and wedded with biotechnology techniques will play a more and more important role in drug development, gene-therapy treatments, and clinical immunology. For example, the Food and Drug Administration has approved nine genetically engineered monoclonal antibodies for treating several diseases, including cancer. Biotechnology will develop human antibodies and other proteins that can be harvested for therapeutic uses. Clinical diagnostics will remain a mainstay of biopharmaceuticals as companies began uniting genomics with microarray (gene-chip) technology to develop new diagnostic tests and ones that can assess an individual's risk of developing a genetic disease.

Microarrays

DNA microarrays will prove an even more essential element in drug development than it currently is as researchers exploit the sequencing of the human genome. These arrays contain thousands of separate DNA sequences on a plate roughly the size of a business card. They enable the analysis of a sample of DNA to determine if it contains polymorphisms or mutations. DNA microassays, which are usually prepared by robotic devices, can be used to diagnose various diseases, identify potential therapeutic targets, and predict toxic effects, and one day they will allow customizing drug regimens for individual patients.

Computer modeling and combinatoral chemistry, which enable the rapid creation of thousands of chemical entities and their testing for potential biological activity, hold the promise of new and safer drugs and, perhaps, lower development costs. Companies are devising and using computer models that assess a compound's absorption, distribution, metabolism, excretion, and toxicity characteristics. The aim is to greatly reduce the number of drugs that go to animal testing and clinical trials, only to prove unacceptable as human therapeutics. Tapping the mapped human genome will reveal many new targets that, combined with computer modeling, will allow drug firms to more rationally design specific drugs, test them at less cost, and feel more confident about them when the companies take them to human trial.

Receptor Activity

Another area of considerable challenge lies in finding ways to control the interactions of hormones and their receptors and other protein-protein relationships. These interactions present potential drug targets, but devising methods to control them will require many technological advances in areas such as synthesis, analysis, computational chemistry, and bioinformatics.

Pharmaceutical companies are also focusing on better ways to deliver drugs, in part as a way to extend their patents. One quest is for better ways to deliver proteins orally. Most proteins used now as therapeutics must be injected, and with a greater number of proteins entering the medical armamentarium in the next 10 years, solving the problems of oral delivery has taken on new urgency. The various drug-delivery methods in development include new inhalation approaches, extended time-release injections, and transdermal administrations. One example of a new transdermal device consists of a tiny pump attached to the skin by an adhesive pad. Pressing a button pushes a needle just below the skin and delivers the drug at a constant rate. Its developers envision that it will be used at first as a way to deliver pain medications.

Agrobiotechnology

Societal pressures may slow innovations in agricultural biotechnology in the next few years, but research will continue, propelled in part by a simple statistic: The world's population is predicted to increase from 6 to 8 billion by 2030. With one-third more humans to feed, people may have no choice but to accept genetically engineered foods.

Among the genetically modified plants currently in fields are insect-resistant corn and herbicide-resistant soybeans, corn, and canola. The potential benefits of genetically modified crops include fewer environmental problems from pesticides and herbicides, increased yields, enhanced nutrition, drought resistance, and even the production of the building blocks of polymers and the remediation of polluted soils, sediments, and aquifers. However, there are unresolved questions about potential risks as well, which have led to opposition to genetically altered foods, particularly in Europe and Japan. These concerns include the possibilities that a gene-altered plant will become an invasive species and cause ecological damage; that plants producing pesticidal proteins may harm nontargeted organisms and/or have an adverse effect on species that feed on the targeted pests; and that new viruses may evolve in virus-resistant plants. Resolving these issues, as well as the lingering controversy over human safety, poses an immediate challenge for researchers and will be necessary for the acceptance of genetically altered plants. In many ways, the future of food produced by biotechnology—at least in the near term—depends on persuading the public of the solid science behind it.

Plant Genomes

Announced last December, the first sequencing of the genome of a higher plant—the weed *Arabidopsis thaliana*—and the nearly completed mapping of the rice genome mark major advances along the way to understanding plant behavior and the opportunities to exploit that knowledge. The rice genome carries particular import for the world's food supply because an estimated 4 billion people will depend on rice as their dietary staple in the year 2030. The *A. thaliana* sequencers predict that the plant has about 25,500 genes, and they have assigned tentative functions to around 70 percent of them. As in the human genome, the complete plant genome enables researchers to compare its DNA sequence with DNA sequences from other plants to identify key genes in cash crops. Plant researchers have set a goal of understanding the function of all plant genes by the end of the decade, which would greatly enhance biotechnologists' ability to generate new genetically altered forms. Those data would then serve as the basis for a virtual plant, a computer model that would enable the simulation of plant growth and development under different environmental conditions.

BLENDING THE PHYSICAL AND THE BIOLOGICAL

For centuries, a sharp demarcation separated the physical and biological sciences, breached only by an occasional discipline such as biophysics. Today, interdisciplinary research is the norm in many industrial laboratories, and not just in the physical or the biological sciences. To a growing degree, research teams may now include representatives from both. Researchers seek to translate biomolecular recognition into useful nanomechanical devices. Geoscientists are exploring genomics for useful clues to solving problems. Cooperative efforts by biologists and engineers seek to create new robots. Some scientists wonder whether deciphering biosignaling in cells will lead to applications in computer science, and others ponder whether the emerging discoveries of brain science will revolutionize information technology. One can expect a greater breaching of the traditional barriers between physical and biological research and a strengthening of biophysical research during this decade—in industry, government, and academic laboratories.

Biotech and Materials Science

One promising area is the interaction of biotechnology and materials science. Biological systems have been used to create two- and three-dimensional inorganic nanoscale structures and assemble gold and semiconductor nanoparticles on a DNA template. Such work aims at goals like nanoscale wires, mechanical devices, and logic elements, as well as creating organic-inorganic compounds. The potential from utilizing the knowledge and skills of biotechnologists and materials scientists includes creation of new molecular switches and transistors, nanosensors, catalytic devices, and opto-electronic components. IBM researchers have demonstrated that molecular recognition between a piece of DNA and its complementary strand can translate into a mechanical response, namely, the bending of a nanocantilever. Researchers trying to develop biocompatible materials for use in human replacement parts draw on new findings from a spectrum of disciplines, including molecular and cellular biology, genetics, polymer and surface science, and organic chemistry.

Robot Engineering

Today, robot building depends almost as much on biologists and neuroscientists as it does on engineers and computer scientists. Robot builders seek insights from the animal kingdom in order to develop machines with the same coordinated control, locomotion, and balance as insects and mammals. The purpose is not to create a robot that looks like a dog (although that has been done and marketed), but to build one—for battlefield use or planet-surface exploration, say—that can walk, creep, run, leap, wheel about, and role over with the same fluid ease as a canine. To do this requires not simply electrical wiring and computer logic, but also a deep understanding of insect and mammalian mobility, which, in turn, requires the input of zoologists, entomologists, and neurophysiologists. What is emerging are some general principles about the complexity of animal mechanics and control. For now, bioinspired robots are mostly creatures of the laboratory. However, one would expect continued development and application of these robots throughout this decade and a backflow of insights to biologists and neurophysiologists as they observe the development of bioinspired machines.

Catalysts

As the number and understanding of enzyme-crystal structures grow, so does interest in utilizing this knowledge to synthesize new catalysts. Researchers envision harnessing such bioinspired catalysts for green chemistry—through the environmentally benign processing of chemicals—and for use both as new drugs and in their production. The effort is in a formative stage, and although chemists have synthesized enzyme-mimicking catalysts, a great deal of attention is focused on deciphering protein structure and its role in enzymatic behavior. Evidence indicates, for example, that an enzyme is not simply a rigid structure on which catalysis occurs, but that the process actively involves the entire protein. A greater understanding of enzymes has exposed the complexity confronting attempts to develop enzyme mimics as well as totally new ones. Questions remain, however, about protein structure and enzymatic mechanisms and how enzymes work in unison within cells. Many answers should emerge in the next few years through biochemistry, protein crystallography, molecule-by-molecule studies of enzymes, and bioinformatics.

Many other examples exist of the melding of the physical and biological sciences. Researchers at Bell Laboratories are trying to exploit both self-assembly and the natural electrochromatic properties of the protein bacteriorhodopsin to develop new flat-panel displays. Unraveling the intricate nature of signaling within cells—the focus of a mini-Human Genome Project called the Alliance for Cellular Signaling—holds clear implications not only for basic biology, clinical medicine, and the pharmaceutical industry but also, potentially, for the computer sciences. Some geoscientists look to the sequencing and analysis of a variety of genomes to aid them in understanding the coevolution of life and Earth and the soft-tissue structure of creatures long extinct. Neuroscientists no longer view the brain as a three-pound biological computer but as an even more complex system that organizes thinking, learning, and memory. Understanding these processes holds significant meaning for computer science and information technology as well.

DISCUSSION AND CONCLUSION

Richard Smalley remarked in a private conversation 6 years ago that the 21st century would be the century of nanoscience. Certainly science at the micro- and nanoscale will play important roles in maintaining the economic competitiveness of the United States in computer chips and storage media, telecommunications, optical devices, display technology, biotechnology, biomedical science, and drug development. At the current time, industry funds more than half of the research most likely to have an economic impact during the coming decade. That situation probably will not change. The current administration appears to view tax policy as a more effective way to stimulate the economy—including its research component—than federal support for research. Although this philosophy might shift with the 2004 or 2008 election, the percentage of research funds provided by the federal government is unlikely to suddenly surge.

Two trends will clearly influence U.S. applied research in science, engineering, and technology in the coming decade. One is the growth of interdisciplinary research. Many projects today require the integration of expertise from several disciplines, team research, and a willingness by scientists and engineers to work with others outside their own discipline to bring them to a successful fruition. The second trend relates to the increasing quality of scientific work emerging from foreign laboratories, particularly in Canada, Europe, and Japan. This work will challenge—as well as inform—U.S. science and technology and may affect its domination in areas such as computers, nanoscience, and biotechnology. Beyond that challenge, international treaties, standards, and regulations will affect U.S. competitiveness. For example, although the United States is not a party to the Cartagena Protocol on Biosafety, its provisions will govern U.S. companies whenever they trade with any country that ratifies it.

The degree to which research will advance during the next 5 to 10 years depends in part on economic, political, and international factors beyond the scope of this report. Although precise predictions cannot be made as to which specific efforts will yield unusually significant results over the next 5 to 10 years, the breadth of the technologies likely to yield extraordinary advances can be identified. However, true breakthroughs, by their nature, are unexpected, unanticipated, and unpredictable.

I

Trends in Industrial R&D Management and Organization

Alden S. Bean

CONTENTS

INTRODUCTION	199
BACKGROUND	200
R&D Trends in the Manufacturing Sector, 201	
Review of the IRI 2008 Forecast, 201	
DISCUSSION OF THE IRI 2008 FORECAST	203
Information Technology, 203	
Globalization of R&D/Technology, 205	
Growing Workforce Diversity, 206	
Integration of Technology Planning and Business Strategy, 208	
Partnerships and Alliances, 211	
Customer Power, 215	
SUMMARY AND CONCLUSIONS	216

INTRODUCTION

This paper examines trends in the organization and management of industrial R&D that may affect the rate and direction of technological innovation over the next 10 years. It builds on a report published by the Industrial Research Institute (IRI) that summarized its members' expectations of how industrial R&D laboratories would operate in 2008.[1] After reviewing findings and projections of that

---

[1] Charles F. Larson, 1998, "Industrial R&D in 2008," Research•Technology•Management 41 (6).

report, more recent trends and projections are discussed. An effort is made to identify "game changers"—changes in structures, policies, or practices that may alter innovation processes or outcomes in significant ways. After a discussion of expected changes in structure and practice across different business sectors, the paper concludes with questions and observations that may merit further discussion.

BACKGROUND

The Industrial Research Institute's forecast of how an industrial R&D laboratory would operate in 2008 is the starting point for this discussion.[2] The forecast resulted from several self-study projects carried out by IRI members between 1996 and 1998, topped off by discussions and presentations at their annual meeting in 1998. For the purposes of this paper, it is important to note both the time frame of these studies as well as the makeup of the IRI membership. As for the time frame, the U.S. economy had been expanding rapidly since 1993; the stock market had been rising steadily since 1982, and from 1992 forward it had risen so rapidly that ". . . the price index looks like a rocket taking off through the top of the chart!"[3] Industrial R&D funding, which had been flat from 1990 to 1994, began to grow at a double digit rate that carried through 2000.[4,5] As for the makeup of the IRI membership, it represents big company R&D, primarily manufacturing firms, with a heavy representation from the chemical sector. The IRI estimates that its 300 member companies perform as much as 85 percent of the nation's R&D. While total industrial R&D in the United States grew almost constantly after the mid-1970s (except for a flat period from 1990 to 1993), manufacturing firms' support for R&D did not grow, in constant dollar terms, from 1986 to 1994.[6] The contribution to total R&D growth between 1986 and 1993 was due to the rise of R&D in the service/nonmanufacturing sector, which is underrepresented in the IRI membership and is made up of much smaller firms.[7]

---

[2]Ibid.

[3]Robert J. Shiller, 2000, Irrational Exuberance, Broadway Books, New York.

[4]Charles F. Larson, 2000, "The Boom in Industry Research," Issues in Science and Technology 16 (4).

[5]Graham R. Mitchell, May 2001, "Industrial R&D Strategy for the 21st Century," in Succeeding in Technological Innovation, Industrial Research Institute, Washington, D.C.

[6]National Science Board, 2000, Science and Engineering Indicators—2000 (NSB-00-1), National Science Foundation, Arlington, Va.

[7]Ibid.

R&D Trends in the Manufacturing Sector

Before discussing the 2008 forecast, it is noteworthy that Fusfeld projected that the principal emphasis for industrial R&D in the 1990s and beyond would be as follows:[8]

- Inside the corporation
  —Coordination of teams with decentralized structures,
  —Convergence of technical strategy with corporate strategy,
  —Establishment of global networks of technical units, and
  —Organization of technical intelligence.
- Outside the corporation
  —Alliances with other corporations,
  —Strategic relationships with universities,
  —Active participation with government programs, and
  —Selective use of consortia and cooperative R&D programs.
- Growth areas
  —Leveraging core R&D with global resources and
  —Expanding technical activity and linkages in developing countries.

Fusfeld cited competitive and cost pressures of the early 1990s as the driving forces for these changes and noted those pressures had "changed the practices of industrial research."

Review of the IRI 2008 Forecast

In Appendix E of this report, McGeary provided a summary of the IRI report by Larson and its conclusions.[9] This discussion expands on McGeary's summary and selects points that deserve amplification or comment in light of more recent events and new information.

First, the report called attention to six forcing functions, identified by IRI members as drivers that would shape the conduct of industrial R&D and innovation in the decade from 1998 to 2008:

- Information technology,
- Globalization of technology,
- Growing workforce diversity,
- Integration of technology planning and business strategy,

---

[8] Herbert I. Fusfeld, July-August 1995, "Industrial Research—Where It's Been, Where It's Going," Research•Technology•Management 38 (4): 52-56.

[9] Charles F. Larson, 2001, "R&D and Innovation in Industry," in Research and Development FY 2001: AAAS Report XXVI, American Association for the Advancement of Science, Washington, D.C.

- Partnerships and alliances, and
- Customer power.

These drivers were subsequently combined in various ways to generate five scenarios for change, which were then projected forward for 10 years and used to characterize what industrial R&D might look like by 2008. The scenarios were the following:

1. *Cyclical change.* This scenario assumes only incremental changes in the way R&D/innovation will be done and that the changes will be cyclical, repeating patterns of change more or less as they occurred in the past. Examples are offered of such patterns, which may take decades to unfold, at GE, DuPont, and Xerox.[10] Hounshell and Smith authored a catalog of R&D management and organization practices at DuPont from 1902 to 1980.[11]

2. *Globalized R&D.* This scenario flows from observations that industrial CEOs (surveyed by the Baldridge National Quality Award Foundation) viewed economic globalization to be the single most important trend affecting U.S. companies and concluded that the trend toward the globalization of R&D would continue or perhaps accelerate through 2008.

3. *R&D through partnerships.* This scenario presumes that technology will become increasingly complex and more expensive to develop, thus forcing companies to rely on partnerships, collaborations, and alliances with other companies, government laboratories, universities, and contract R&D organizations to build the critical mass of skills and know-how needed for innovation.

4. *Innovation function absorbs R&D.* This scenario assumes that the innovation (new products, processes, services) value proposition will overwhelm the perceived need for a separate R&D function, thus resulting in complete decentralization of R&D into customer-focused business units. This amounts to a complete and permanent victory for the decentralized model of R&D organization, the pros and cons of which are thoroughly described by Hounshell and Smith.[12]

5. *Networking counts.* This scenario acknowledges the well-documented fact that "invisible colleges" and networks are endemic to science and technology, and that increased recognition of ways to expand and extract value from networks will change traditional R&D. (If Metcalfe's law and network externalities facilitate the trend toward expanding the scope of "networking" to new content areas, then perhaps security concerns and firewalls will constrain it.)

---

[10]Larson, 1998, op. cit.
[11]David A. Hounshell and John Kenly Smith, Jr., 1988, Science and Corporate Strategy, Cambridge University Press, New York.
[12]Ibid.

Larson concludes by stating his belief that the most likely scenario will be some hybrid of the five.[13] He goes on to describe the four major differences between the laboratories of 1998 and 2008:[14] people, technology intelligence, technical work, and the innovation process.

DISCUSSION OF THE IRI 2008 FORECAST

By identifying information technology and globalization as the most important of the six drivers of change in R&D organization and management, Larson's *Industrial R&D in 2008* invites closer examination of these forces and how they will play out as the scenarios unfold. Will they be "game changers" or evolutionary enhancements? Here are some further thoughts and observations.

Information Technology

Information technology (IT) is a broad category that includes computers, data networks, the Internet, and the World Wide Web but can also include telecommunications. Shiller considers the arrival of the Internet and the World Wide Web beginning in 1993 to have been one of the 12 factors upon which the (irrationally exuberant) bull market of the 1990s was based.[15] On the other hand, Levine et al. propose that the Internet and the Web will change business as usual and work wondrous transformations on business structures and processes.[16] A recent paper by Litan and Rivlin, which analyzes the likely impact of the Internet on the U.S. economy over the next several years, concludes that "there are many reasons to side with the optimists, especially those who claim that widening use of the Internet creates a significant potential for increasing productivity and increasing the standard of living over time."[17] The paper cites four major findings regarding pathways for these effects:

- Reducing transaction costs in financial services, health care, government, education, and manufacturing;
- Increasing the efficiency of management in automobile manufacturing, trucking operations, and supply chain management;

---

[13]Larson, 1998, op. cit.

[14]Michael McGeary, May 2001, Recent Reports on Future Trends in Science and Technology: A Background Paper for the NRC Committee on Future Environments for the National Institute of Standards and Technology, Washington, D.C.

[15]Shiller, 2000, op. cit.

[16]Rick Levine, Christopher Locke, Doc Searles, and David Weinberg, 2000, The Cluetrain Manifesto: The End of Business As Usual, Perseus Publishing, Cambridge, Mass.

[17]Robert E. Litan and Alice M. Rivlin, 2000, The Economy and the Internet: What Lies Ahead, Internet Policy Institute, Washington, D.C.

- Making markets more compatible by lowering information costs and frictions; and
- Increasing consumer choice and convenience.

The Litan and Rivlin paper concludes that the potential of the Internet to contribute to productivity growth is real (although noting that a 2 percent annual growth rate is more likely than the 3 percent experienced between 1995 and 2000); that the gains due to e-commerce are likely to be less important than the reductions in transaction costs in financial and health services, government, and old economy firms in general; that enhanced management efficiencies will increase productivity in product development, supply chain management, and many other aspects of business; and that we should not overestimate the productivity-enhancing effects of the expected improvements in market competitiveness, because much of the cost improvement will be passed along to consumers as price reductions.[18]

How important might IT be to the future of R&D? The members of the industrial R&D community also seem to have come down on the side of the optimists. Setting aside the slowdown in demand for IT equipment, the build-out of the IT infrastructure, and the dot-com e-commerce disappointments, there is evidence that the Internet and the World Wide Web are enabling important changes in the way R&D and innovation are managed. If we remember that the Web was invented by the international community of physicists to improve their ability to communicate and collaborate in the design of the particle physics research program carried out at CERN in Switzerland, this should not be a total surprise.

In 1999, Paul Horn, senior vice president for research at IBM, wrote "Information Technology Will Change Everything," in which he projected key improvements in IT through 2025.[19] He captured the primary impacts of these technical gains in three concepts: (1) "the wired planet" depicts the globalization of the Internet—"everyone and everything" becomes connected to the Internet; (2) "knowledge becomes the new currency"—with an estimated 300 million home pages on the Web, and with new pages being added at the rate of 800,000 a day, how do you find the pages that contain what you need to know; and (3) "augmented reality"—a world in which computers will be able to reason and interact with human beings very much like we interact with one another. Horn concluded his paper by challenging managers to think about three implications of IT growth for their companies:

- What will your brand identity mean in a world where everyone can get what they need on the Internet?

---

[18]Ibid.
[19]Paul M. Horn, 1999, "Information Technology Will Change Everything," Research•Technology•Management 4 (1): 42-47.

- Are your systems ready for your company to be an Internet company for knowledge management?
- Are your people ready to function in a networked world?

Perhaps because of Horn's challenging questions, industrial R&D managers have continued to confer about and publish on these issues. Browsing past issues of the Industrial Research Institute's journal, *Research •Technology•Management* (*RTM*), turns up articles on knowledge management, organizational structure, globalization, and intellectual property that provide insights about the ways IT is changing R&D practice. Some of the ways mentioned include the following:

- Leveraging knowledge management toward the goals of total quality management (TQM) and total customer satisfaction;
- Leveraging cross-company resources and permiting rapid integration of new business acquisitions;
- Flattening organizations and defeating command and control management styles; and
- Enabling virtual research teams to transcend geographic, linguistic, and temporal barriers.

Globalization of R&D/Technology

The Commerce Department reported that R&D by foreign firms in the United States grew throughout the 1990s as much as R&D by U.S. firms located abroad.[20] These trends are expected to continue. The motivation for overseas R&D continues to be the same as it was in the early 1990s: assisting parent companies to meet customer needs, keeping abreast of technological developments, employing foreign scientists and engineers, and cooperating with foreign research laboratories.[21]

The IRI has begun to feature news of international R&D activities in *RTM*. The January-February 2001 issue detailed current activities in China, Sweden, and Russia, while the March-April 2001 issue covered the Asia-Pacific region. These articles highlighted domestic R&D activities of those countries and also covered the activities of U.S. firms that have laboratories or cooperative programs in those countries. The impression gleaned from these articles is that barriers to international knowledge flow are coming down, thanks to IT; that intellectual property policy regimes are moving in the direction of being inter-

---

[20]Donald H. Dalton, Manuel G. Serapio, and Phyllis Genther Yoshida, September 1999, Globalizing Industrial Research and Development, U.S. Department of Commerce, Technology Administration, Office of Technology Policy, Washington, D.C.

[21]Ibid.

nationalized; that strategic alliances are increasingly popular; and that venture capital support for entrepreneurial activity is increasing throughout the world.

Quantitative data from the National Science Board (NSB) show the globalization of R&D as well as the underlying composition of R&D expenditures.[22] Industrial support for R&D has been rising relative to government support throughout the G7 countries, averaging 62 percent of total funding across OECD countries in 1997. R&D funded by foreign sources in these countries doubled between 1981 and 1997, but it still accounted for only 2.5 percent of all R&D funds. While there is substantial variation in industry/government funding ratios across the G7 countries (from a high of 74 percent by industry in Japan to a low of 30 percent in Russia), the performer base is decidedly industrial, ranging from about 70 percent in the United States to about 54 percent in Italy. Academia is the second most important performer of research, with basic research the dominant activity. International cooperative R&D and strategic technical alliances have doubled since 1980, with IT and biotechnology alliances more important than all others combined.[23]

Growing Workforce Diversity

In the early 1990s, concerns over R&D workforce diversity referred to management's need to deal effectively with gender and ethnic differences among scientists and engineers entering the workforce. The issues had to do largely with their enculturation and assimilation into the role structure of laboratories and related functions. Today and for the foreseeable future, industry will be concerned with these same issues, but in an expanding context. It is concerned not just with R&D, but with the full set of roles and behaviors required to make industrial innovation happen rapidly and successfully. The knowledge management emphasis (discussed above), together with the strategic management perspective (discussed below), is driving industry to place a high value on recruiting and retaining people with strong credentials in scientific and engineering fields where explicit knowledge represents a core technical competency; on retaining people whose tacit knowledge is critical to successful innovation based on core technical competitiveness; and on outsourcing its noncore needs. Essentially, a value chain/supply chain management mentality is being applied to the R&D/innovation management process.

An article by Perry illustrates the retention problem, which peaked during the dot-com craze of the late 1990s but which "won't be over for a long time because of a shortage of talent," according to an interviewee at Corning.[24] Perry sums up the problem thusly:

---

[22]National Science Board, 2000, op. cit.
[23]Ibid.

As dot-com mania fades, its effect on job mobility lingers on. Star technical performers, once considered organizational loyalists, are more likely to jump ship for competing job offers. Employers are responding to the threat by introducing new policies and procedures designed to retain key personnel. The most visible and expected change is in compensation: Salary, equity participation and bonuses are on the upswing, as is the correlation of pay with performance and the frequency for which compensation is reviewed. But good pay is not enough to keep the best people in a new environment characterized by talent raids by competing organizations. Star performers are demanding, and getting, more opportunities for career advancement, greater participation in leading-edge projects, and entrepreneurial freedom.[25]

A series of articles has appeared in *RTM* dealing with these new diversity issues. Demers writes as follows:

The type of talent organizations most fear losing are typically people in:

- Technology who are implementing e-commerce strategies,
- Information technology who are developing and/or implementing major new systems,
- Junior positions where people are slated for major leadership roles,
- Sales to customers considered vital to the preservation of revenues,
- Positions with unique knowledge of products and services, and
- Positions needed to support a major consolidation or acquisition.[26]

Other papers deal with defining the specialized roles required to make innovation happen—for example, Markham and Smith on the role of champions in the new product development process[27] and Roberts on the critical executive roles required for strategic leadership in technology.[28] As innovation continued to absorb more and more of the attention and resources of industrial firms, as compared with traditional R&D laboratory work, large industrial firms have had to deal with "entrepreneurship" in more creative ways, as Perry points out.[29] In many cases this has resulted in changes in organizational structure, including the

---

[24]Phillip M. Perry, 2001, "Holding Your Top Talent," Research•Technology•Management 44 (3): 26-30.

[25]Ibid.

[26]Fred Demers, 2001, "Holding On to Your Best People," Research•Technology•Management 44 (1): 13-15.

[27]Stephen K. Markham and Lynda Aiman-Smith, 2001, "Product Champions: Truths, Myths and Management," Research•Technology•Management 44 (3): 44-50.

[28]Edward B. Roberts, 2001, "Benchmarking Global Strategic Management of Technology," Research•Technology•Management 44 (2): 25-36.

[29]Perry, 2001, op. cit.

creation of internal venture capital groups in several firms and the augmentation of the laboratory system to include the equivalent of incubators. (See, for example, the discussion of Xerox's newly formed Xerox Technology Enterprise division in Lautfy and Belkhir.[30]) These structural changes are interesting in their own right, but their staffing patterns are also noteworthy because of the academic credentials of the principals. For example, both of the senior managers of Xerox XTE division hold Ph.D.'s in physics and appear to have been very productive scientists. Both have been drawn out of traditional R&D roles into these new innovation management responsibilities.

The National Science Board acknowledges the increasing complexity of the labor market for professional scientists and engineers and notes that changes in the structure of career opportunities in science and engineering (S&E) need to be carefully matched to career expectations and training. Otherwise, the health of scientific research in the United States could be damaged, which could "reduce the ability of industry, academia and the government to perform R&D, transfer knowledge, or perform many of the other functions of scientists in the modern economy."[31] Since industry employs such a high percentage of the nation's S&E professionals (77 percent of B.S. degree holders, 60 percent of M.S. degree holders, and 38 percent of all Ph.D.'s), shifts in the structure of industrial R&D/innovation work could have important implications for S&E education programs.

Integration of Technology Planning and Business Strategy

In retrospect, the recession of 1990 and 1991 seems to have catalyzed serious corporate efforts to rationalize the relationship among business strategies, organizational structures, and operating systems. There was great concern over the Japanese "miracle" in manufacturing, and a flurry of business books appeared that urged corporate reforms. The relationship between business strategy and R&D operations was not a new topic by any means. (See Hounshell and Smith for an account of DuPont's struggle with the problem between 1902 and 1980.[32]) A book by three Arthur D. Little, Inc., consultants drew attention to the technology planning/business strategy issue in a readable and persuasive way.[33] It undoubt-

---

[30]Rafik Lautfy and Lotfi Belkhir, 2001, "Managing Innovation at Xerox," Research•Technology•Management 44 (4): 15-24.

[31]Steven W. Popper and Caroline S. Wagner, 2001, New Foundation for Growth: the U.S. Innovation System Today & Tomorrow, An Executive Summary, MR-1338.0/1-OSTP, RAND Science and Technology Policy Institute, Washington, D.C.

[32]Hounshell and Smith, 1988, op. cit.

[33]Philip A. Roussel, Kamal N. Saad, and Tamara J. Erickson, 1991, Third Generation R&D: Managing the Link to Corporate Strategy, Harvard Business School Press, Boston, Mass.

edly served as a focal point for many corporate discussions and decisions that have reshaped business/technology strategy linkages.

Since the information requirements for developing good business strategies and technology roadmaps are very demanding, and much of the desired information is proprietary and/or unpublished, industry has spent a great deal of time and money on competitive intelligence and data-mining in recent years. Improved databases and software for extracting valued information from data (as discussed above) are making the construction of business/technology portfolios and roadmaps more feasible year by year. As these costs come down, strategy formulation will perhaps become less challenging than strategy implementation and execution.

Surveys in 1992 and repeated in 1999 provide insights into trends in global strategic management of technology practices. Roberts reports on 209 responses received from 400 very large firms surveyed in the United States, Japan, and Europe, and presents a preliminary interpretation of the results.[34] Here are some of the observations:

- The integration of corporate technology strategy and overall business strategy depends heavily on the involvement of senior management in formulating and implementing strategy. Trends are as follows:
    —CEOs, R&D VPs, and CTOs are the primary integrators, with marketing VPs and CFOs lagging behind.
    —Japanese and North American firms see themselves as having strongly linked business/technology strategies, while Europe lags. The trend was upward for Japan and North America between 1992 and 1999.
    —96 percent of Japanese firms placed R&D executives on their boards of directors, compared with 35 percent in Europe and 8 percent in North America.
    —91 percent of Japanese, 67 percent of European, and 60 percent of North American companies have CTOs or R&D VPs on their executive committees.
- Those firms with the strongest linkages between corporate business/technology strategies are the strongest business performers on the following measures:
    —Overall corporate sales growth rates,
    —Share of 1998 sales from new products and services,
    —Share of sales from new and improved products,
    —Technology leadership, and
    —Perceived R&D timeliness in meeting new product delivery schedules.
- Between 45 and 49 percent of the firms surveyed in the three regions had CEOs with strong technical backgrounds.

---

[34]Roberts, 2001, op. cit.

- The greater the percentage of the total R&D budget spent on short-term R&D, the better the R&D performance on the following measures:
  —Perceived efficiency,
  —Timeliness,
  —Improved time to market,
  —Meeting market dates for commercialization, and
  —Process implementation.
- Except for the pharmaceuticals industry, spending on research (basic and applied) declined relative to spending on development and support services for both corporate and divisional R&D between 1991 and 1998.
- Globalization of the R&D performer base increased for firms in all three regions.
- External sourcing of technology increased substantially between 1992 and 1998 and was expected to increase even more by 2001:
  —The percentage of firms relying on external sources of technology rose from about 20 percent to about 70 percent by 1998 and is expected to rise above 80 percent by 2001.
  —Central corporate research is the most important source of research, despite the trend toward external technology acquisition.
- Regarding overall technical effectiveness:
  —The greater the technical maturity of a business unit's key technologies, the lower the performance of the business.
  —The greater the company's technical leadership position, the stronger its performance.
  —Speed to market is associated with many possible influences, including more short-term R&D.

Roberts also identified improved technology management practices in the following areas as important:

- Use of automation and other nonhuman resources in R&D,
- Use of IT and communications technology, and
- Streamlined R&D organizations.

In his conclusions, Roberts emphasized that

- Strong business/technology strategy linkages are correlated with superior business performance.
- "The most important continuing business change in strategic technology management is the increasing worldwide reliance on external-to-each-company sources of technology."

- The emphasis on speed to market has "reinforced the disproportionate emphasis on near-term R&D spending at the expense of longer-term strategies."

Before leaving this topic, it is interesting to note the elements of a presentation by Edelheit at which he spoke on new pathways of innovation at General Electric.[35] After noting that GE had been transformed from a company that derived 85 percent of its revenue from manufactured products in 1980 to one that derives 65 percent of its revenues from services today, he noted the following changes in the role of corporate R&D that are going on today:

- Shifts in funding from corporate to business units;
- Project management a major focus;
- Short-term technical support part of mission;
- Synergy across businesses;
- Ensure leadership technology in the businesses;
- R&D time frame shortened;
- "Game changers" still a priority;
- Multigeneration product development;
- Partnerships and joint ventures a way of life;
- Supporting corporate initiatives—for example, six sigma and e-engineering; and
- Human resources strategy.

The overlap between these changes at GE and those reported by Robert is striking and perhaps a sign that these issues will drive corporate technology strategies for some time to come.

Partnerships and Alliances

Industrial R&D partnerships and alliances appear to be motivated by a variety of corporate concerns and circumstances. The NSB broadly characterizes R&D partnerships as collaborations that allow "individual partners to reduce costs and risks and enabling research ventures that might not have been undertaken otherwise."[36] Moreover, "the underlying theme is that more can be accomplished at lower costs when resources are pooled, especially if organizations can compliment each other in terms of expertise and/or research facilities." The report goes on to detail special conditions surrounding alliances involving different types of partners—industry to industry, industry-academic, industry-federal laboratories,

---

[35]Lewis S. Edelheit, 2001, "New Pathways to Innovation at GE Corporate R&D," presented at the Industrial Research Institute Annual Meeting, May 20-23.

[36]R. Mitchell, 2001, op. cit.

academic-federal government—and to briefly describe the impact of international competition on government policy making decisions regarding the formation of R&D consortia.

After reading the results of Roberts' survey and discussions of the use of R&D alliances in industry by Wyndrum,[37] Campbell,[38] Brenner and Tao,[39] and Miller and Morris,[40] it is difficult to say whether the increase in external sourcing of R&D should be expected to continue for the foreseeable future or whether it is a more transient phenomenon. To be specific, despite the federal laws enacted since 1980 to promote industry-government partnerships and to promote technology transfer, the number of both corporate research and development agreements (CRADAs), which involve industry/government laboratory collaborations, and industry/industry research joint ventures (RJVs) registered with the Department of Justice seems to have peaked in 1996 and is now in decline.[41] Additionally, federal funding for NIST's Advanced Technology Program (ATP) has been a political football, having suffered a major cutback in 1996, and is widely expected to be terminated by 2002.

Moving to research collaborations within industry, the pharmaceutical industry is noted for its use of combinations of in-house R&D, research contracts to small external research labs, and merger and acquisition activity to sustain innovation across successive waves of technological change.[42] Campbell writes:

> The explosion in new technologies and genome sciences has created a unique situation for the pharmaceutical industry in that no single company has the resources or expertise to pursue all of these innovative techniques in-house. Thus outsourcing is becoming much more common, with some companies investing up to 20 percent of their research budgets in external collaborations.[43]

In addition to small research firms, universities are important R&D collaborators for industrial firms. As noted earlier, universities have overtaken the federal

---

[37]Ralph Wyndrum, Jr., 2000, "A New Business Model for R&D: Acquisitions and Minority Equity Investments," unpublished presentation to the National Technological University, Vail, Colo., August.

[38]Simon F. Campbell, 2001, "Individuals, Teams and Innovation in the Pharmaceutical Industry," Succeeding in Technological Innovation, Industrial Research Institute, Washington, D.C.

[39]Merrill S. Brenner and John C. Tao, 2001, "You Can Measure External Research Programs," Research•Technology•Management 44 (3): 14-17.

[40]William L. Miller and Langdon Morris, 1999, Fourth Generation R&D: Managing Knowledge, Technology, and Innovation, John Wiley & Sons, New York.

[41]National Science Board, 2000, op. cit.

[42]Basil Achilladelis and Nicholas Antonakis, 2001, "The Dynamics of Technological Innovation: the Case of the Pharmaceutical Industry," Research Policy, April, pp. 535-588; reviewed in Research•Technology•Management 44 (4): 63.

[43]Campbell, 2001, op. cit.

government as the second most important performer of R&D. A recent paper by Hicks et al. notes that patenting activity in IT and health technologies is growing rapidly in the United States and that university patenting is particularly active.[44] It further notes that company patents frequently cite publications from "in-state public sector institutions," perhaps indicating a diffusion of expertise beyond the elite universities.

In the chemical industry, Brenner and Tao report that Air Products and Chemicals has been involved in 42 external R&D partnerships since the mid-1990s. Air Products has saved 2 years of internal effort and hundreds of thousands of dollars in net research expenditures per project and has the potential to generate tens of millions of dollars per year in increased revenues and millions in profits. The company's partners included industrial consortia, universities, government labs, and unique (often small) research and development companies.[45]

With regard to the telecommunications sector, the complex interplay between the various service and carrier companies across wireless, cable, land lines, and Internet technologies has stimulated major restructuring efforts at large, established telecoms and Internet service providers as well as at smaller newcomers. Wyndrum conducted a comprehensive analysis of technology investment activities at 28 of these firms between 1998 and 2000.[46] In particular, he studied the motives behind their outright acquisitions and minority equity investment (MEI) activities. Three motives were grouped together and called R&D: acquiring access to (1) technology for new products, (2) technology for new features or functionality of existing products, (3) and skills. The other motive studied was market expansion. Wyndrum found transactions worth more than $300 billion, of which $15 billion was in the form of MEIs and the balance was for direct acquisitions. Upon further analysis, he found that firms that lacked internal R&D spent $274 billion to acquire firms for market expansion, $12.9 billion to acquire R&D expertise, and $3.3 billion to take MEI positions, of which $0.9 billion went for market expansion and $2.3 billion went to acquire R&D capability. On the other hand, he found that firms with strong internal R&D of their own spent $10.2 billion for acquisitions, of which $4.5 billion was for market expansion and $5.7 billion for R&D capability; of the $11.9 billion in MEIs, $5.3 billion was for market expansion and $6.3 billion for R&D expertise.

Finally, Wyndrum notes that the net result of these MEI/acquisition activities was to move firms with little or no internal R&D to a position comparable to

---

[44]Diana Hicks, Tony Breitzman, Dominic Olivastro, and Kimberly Hamilton, April 2001, "The Changing Composition of Innovative Activity in the U.S.—A Portrait Based on Patent Analysis," Research Policy, pp. 681-703; reviewed in Research Technology Management 44 (4): 63.

[45]Brenner and Tao, 2001, op. cit.

[46]Wyndrum, 2000, op. cit.

established firms with strong internal R&D (between 8 and 10 percent of revenues) and that companies that already had strong internal R&D maintained or strengthened their R&D/sales ratios as a result of the MEI/acquisition investments. Thus, to the extent that external R&D collaborations evolve into joint business ventures and eventually into merger and acquisition activities, industry restructuring is facilitated.

An article by Germeraad calls attention to other practices in external R&D/technology development that may be important emerging trends.[47] His basic point is that firms have done such a good job of integrating their technology strategies and business strategies that they are finding new ways of appropriating returns from their R&D investments. Germeraad describes these methods as follows:

- *New avenues for building corporate value.* These trends encompass new avenues to building corporate value. These include the changing basis of market capitalization, building companies to flip, off-balance-sheet spin-offs, and licensing-out as a business practice. These market forces are redefining what value means to a company. The basis of market capitalization is changing from a focus on tangible to intangible assets. Companies are no longer being built solely to withstand the tests of time. Instead, some new companies are being built to flip. Spin-offs are going off the balance sheet and licensing-out is emerging as its own business unit. All of these trends are affecting how a CEO explains value to the company's shareholders and how R&D expenses are explained to the analysts.
- *New development processes.* Second, new development processes afford tremendous opportunity for substantial returns from the new business models. These trends are "spiral" product development, synchronous portfolio selection and management, iterative integration of product development and portfolio management, and connection-focused new product developments. Product development and portfolio selection are moving from conventional funnel models to spiral development. Acquisitions and partnerships are replacing some of R&D's traditional role in bringing new products to the company. These trends are both reducing the time to bring new products to market and adding greatly to the complexity of managing R&D.
- *Trends that are warping time.* The third group of trends relates to "warping time." These are enabling trends that are improving the productivity of R&D, yet these new capabilities are so far neither robust nor embedded in R&D organizations. They include accessible venture funding, the advent of the image age, and the amalgamation of R&D and general management methods. Readily available venture capital funding is accelerating the development of new business opportunities. The ability to view massive amounts of structured and unstructured

---

[47]Paul Germeraad, 2001, "The Changing Role of R&D: Ten Trends That Shape Today's Product Development," Research•Technology•Management 44 (2): 15-20.

information as interactive pictures is enabling better business decisions at high speed. Collectively, all of the "programs of the year" have, in fact, improved the productivity of R&D organizations and increased the speed and quality of product development. Yet, if not managed carefully, these trends introduce higher risks associated with near-synchronous and/or overlapping project and portfolio activities.

To summarize, economic globalization as well as scientific and technological breakthroughs disrupt traditional sources and means of innovation. To the extent that these forces continue in place over the next 10 years, R&D partnerships and collaborations can be expected to continue, as part of the dynamic that restructures and expands industry.

Customer Power

Chatterji and Davidson credit the total quality management (TQM) movement not only with making U.S. R&D more customer-focused but also with bringing R&D into closer contact with the rest of the business enterprise and making it more business-oriented than in the past.[48] The TQM movement began in Japan in the 1970s and 1980s and picked up momentum in the United States in the late 1980s and early 1990s, when it began to impact the R&D function. The primary tenets of TQM are as follows:

1. Meet (and if possible, exceed) the customer's needs and expectations in all areas of contact.
2. Focus on the process that produces a product or an output, not just on the product itself, and in so doing, drive out inefficiencies and waste in all business processes.
3. Incorporate improvements in the organization at all levels, and continuously strive for additional improvement opportunities.

Chatterji and Davidson say that TQM, when applied to R&D, introduced many reforms into R&D management in the mid-1990s, including breaking down the not-invented-here barrier, resulting in increased outsourcing of R&D and acceptance by R&D professionals of performance metrics.

Similarly, the movement to reduce R&D/innovation cycle time is considered to be customer driven.[49] Customization of the SONY Walkman based on fast

---

[48]Deb Chatterji and Jeffrey M. Davidson, 2001, "Examining TQM's Legacies for R&D," Research•Technology•Management 44 (1): 10-12.

[49]Ibid.

response to customer feedback was so successful that it stimulated changes in U.S. innovation management practices, including concurrent engineering, cross-functional teams, and the widespread acceptance of the product platform concept.[50]

Perhaps the strongest influence of the "closer to the customer" movement in U.S. industrial R&D has been the apparent decline in the importance of the corporate/central R&D lab in the United States. This trend, noted by Roberts,[51] Hicks,[52] and others in addition to Chatterji and Davidson,[53] has resulted in shorter time frames for industrial R&D, as noted earlier. Moreover, as shown by Hicks et al., during the decline of the corporate lab, the rate of industrial patenting slowed relative to university patenting, and in all technologies except health, the linkage to science (as measured by citations to scientific literature) is much stronger in patents issued to universities than in those issued to industry. (In health, the science connection is stronger in both university-owned and government-owned patents, but the rates of change are closer than in other technologies.[54])

Thus, while the nation is now more dependent on industrial R&D for its technological strength than it has been since before World War II, industrial invention appears to be less science-based than it was 20 years ago. While the pickup in university patenting and its strong linkage to science may suggest that universities may be offsetting the trend in industry, Hicks et al. caution that there is evidence that the surge in university patenting may reflect the pursuit of quantity at the expense of quality.[55]

SUMMARY AND CONCLUSIONS

It is clear from the literature reviewed above that industrial R&D managers believe that important changes are taking place in R&D management practices and structures, and that they expect gains in R&D/innovation efficiency and effectiveness to be realized as a result. Analysts seem to agree that interesting changes are taking place but are at odds about their expected impacts. The crux of the disagreement is over productivity growth effects. Gordon, in a recent paper published by the National Bureau of Economic Research,[56] is skeptical of future

---

[50]Marc H. Meyer and Paul C. Mugge, 2001, "Making Platform Innovation Drive Enterprise Innovation," Research Technology Management 44 (1): 25-39.
[51]Roberts, 2001, op. cit.
[52]Hicks et al., 2001, op. cit.
[53]Chatterji and Davidson, 2001, op. cit.
[54]Hicks et al., 2001, op. cit.
[55] Ibid.
[56]Robert Gordon, 2000, "Does the 'New Economy' Measure Up to the Great Inventions of the Past?" Working Paper #7833, National Bureau of Economic Research, Cambridge, Mass.

productivity gains, because the innovations we are counting on for those gains, mainly computer, telecommunications, and Internet technologies, are simply not sufficiently robust to sustain productivity growth rates comparable to those realized from the technologies that we produced in the late 1800s and early 1900s (electricity, internal combustion engines, the chemical and pharmaceutical industries, the entertainment, information, and communication industries, and urban sanitation infrastructures). Gordon regards modern innovations to be primarily substitutes for older technologies, with little upside potential. Similarly, Shiller feels that the Internet pales in comparison with the interstate highway system in its productivity-enhancing potential. He regards the Internet and the World Wide Web as part of the new economy hype that created the stock market bubble of the 1990s and little else.[57] Schwartz published an article that presented three future scenarios for "recovery": one is V-shaped, a second is U-shaped, and the third is L- shaped, which is not a recovery at all! As discussed earlier, Litan and Rivlin are more optimistic about the productivity-enhancing potential of the Internet, but cautiously so.[58] Thus, to the extent that changes in R&D management practices and structures are Internet/Web driven, analysts would probably caution against being too optimistic about these long-term social and economic impacts.

The trends of note are the following:

• A decided shift away from corporate funding toward business unit funding for R&D is under way. While many large corporations (GE is one) have decided to use a hybrid structure for their laboratory system (a combination of corporate and divisional business unit labs), corporate funding of corporate labs has dried up. As a result, industrial R&D is more short-term-oriented and under great pressure to meet the time to market and other immediate needs of the business units that now provide most of their funds.

• The growth of service sector R&D from 5 percent of total industrial R&D in 1983 to 26 percent by 1993 has stabilized at 22 percent. The NSB notes the difficulty of disaggregating these numbers into specific sectors, making it difficult to tell what kind of R&D is being emphasized.[59] Moreover, the articles by Edelheit[60] and Campbell,[61] which discuss respectively the transformation of GE into a service company and the expenditure of 20 percent of some pharmaceutical firms' budgets on external R&D, suggest the need for better data so trends can be better discerned and understood.

---

[57]Shiller, 2000, op. cit.
[58]Litan and Rivlin, 2000, op. cit.
[59]National Science Board, 2000, op. cit.
[60]Edelheit, 2001, op. cit.
[61]Campbell, 2001, op. cit.

- The rise of university patenting propensity relative to that of industry (in fields other than health) reflects the increasing desire of universities to participate in the intellectual property marketplace. The declining number of references to scientific publications in university patents suggests a preference for quantity over quality.
- A trend toward involving professional scientists and engineers in all facets of the technological innovation process, not just R&D, raises questions about the training of young scientists and engineers for entry-level positions in industry.
- The external sourcing of R&D noted by Roberts, as well as the emerging trends in creating/capturing value from R&D results noted by Germeraad, deserves tracking.[62,63] The interaction of these activities with existing tax policies and financial accounting standards seem worth watching, as firms continue to find ways to include intellectual capital on the balance sheet.
- Globalization will interact with the Internet and the World Wide Web to create hybrid corporate structures and relationships that we are just beginning to understand. R&D networking will increase as Metcalfe's law operates within specific scientific and technical communities. Mobility of S&E employees will make knowledge management increasingly difficult, and great pressure will emerge to standardize and enforce intellectual property rights internationally. Universities and industry will struggle more and more with intellectual property rights conflicts as the stakes get higher.

---

[62]Roberts, 2001, op. cit.
[63]Germeraad, 2001, op. cit.